Radiotherapy in practice: brachytherapy

D0757126

Radiotherapy in practice

Forthcoming volumes in the series:

Radiotherapy in practice: external beam therapy
Peter J. Hoskin

Radiotherapy in practice: radioisotope therapy
Peter J. Hoskin

Radiotherapy in practice: brachytherapy

Edited by

Peter Hoskin

Consultant Clinical Oncologist
Mount Vernon Hospital, Northwood, Middlesex, UK;
Professor of Clinical Oncology
Royal Free and University College London Medical School,
London, UK

and

Catherine Coyle

Consultant Clinical Oncologist
Cookridge Hospital, Leeds, UK

OXFORD
UNIVERSITY PRESS

OXFORD

UNIVERSITY PRESS

Great Clarendon Street, Oxford OX2 6DP

Oxford University Press is a department of the University of Oxford.
It furthers the University's objective of excellence in research, scholarship,
and education by publishing worldwide in

Oxford New York

Auckland Bangkok Buenos Aires Cape Town Chennai
Dar es Salaam Delhi Hong Kong Istanbul Karachi Kolkata
Kuala Lumpur Madrid Melbourne Mexico City Mumbai Nairobi
São Paulo Shanghai Taipei Tokyo Toronto

Oxford is a registered trade mark of Oxford University Press
in the UK and in certain other countries

Published in the United States
by Oxford University Press Inc., New York

© Oxford University Press 2005

The moral rights of the author have been asserted

Database right Oxford University Press (maker)

First published 2005

All rights reserved. No part of this publication may be reproduced,
stored in a retrieval system, or transmitted, in any form or by any means,
without the prior permission in writing of Oxford University Press,
or as expressly permitted by law, or under terms agreed with the appropriate
reprographics rights organization. Enquiries concerning reproduction
outside the scope of the above should be sent to the Rights Department,
Oxford University Press, at the address above

You must not circulate this book in any other binding or cover
and you must impose this same condition on any acquirer

A catalogue record for this title is available from the British Library

Library of Congress Cataloging in Publication Data
(Data available)

ISBN 0 19 852940 6 (Pbk)

10 9 8 7 6 5 4 3 2 1

Typeset by Cepha Imaging Private Ltd., Bangalore, India.
Printed in Great Britain
on acid-free paper by Biddles Ltd, King's Lynn

Foreword

C.A.F. Joslin,

Emeritus Professor of Radiotherapy,

The University of Leeds.

Until the 1960s the major disadvantage of brachytherapy was the radiation hazard to staff, other patients, and patients' visitors. From 1960 onwards, remotely controlled afterloading techniques either at low or high radiation dose rates became a practical possibility. The available nuclides were then greatly restricted, as were the methods of treatment application.

Since that time, engineering and scientific developments have provided the clinician with the opportunity to develop and introduce novel brachytherapy procedures to almost any body site where brachytherapy is considered to offer advantages against other treatment methods.

Publications abound in the various journals that deal with the physics, radiobiological, and clinical issues. However, there are few text books which cover all aspects of brachytherapy that apply not only to trainees in physics, radiobiology, or radiation oncology but also to those practitioners at the expert level.

The authors of this book recognize the fact that no single book can cover all aspects of the subject and a comprehensive coverage can be achieved by reading the commended supportive reading from the literature. They have taken a practical approach in conjunction with the scientific fundamentals of the subject. The authors are among the leading brachytherapy experts in the United Kingdom and have dealt with the subject matter with clarity and exposition.

It has been a long haul since I was fortunate enough to have been involved in the development and application of high dose rate brachytherapy in the 1960s using a Cathetron machine. This was made possible due to funding from the Department of Health as part of a Research and Development project. Although perhaps biased, I consider this has proven a good investment for the Ministry and that support may continue to be provided to enable this important work of progressing further the role of brachytherapy in the treatment of patients with cancer. The authors are to be congratulated on the important part they also play in this development by the publication of this book.

Acknowledgements

We are indebted to the Medical illustration departments at Cookridge Hospital Leeds and Mount Vernon Hospital, Northwood.

Varian Brachytherapy kindly supplied the following figures: Fig. 5.5(A), Fig. 8.1, and Fig. 9.1(A).

Oxford University Press makes no representation, express or implied, that the drug dosages in this book are correct. Readers must therefore always check the product information and clinical procedures with the most up to date published product information and data sheets provided by the manufacturers and the most recent codes of conduct and safety regulations. The authors and the publishers do not accept responsibility or legal liability for any errors in the text or for the misuse or misapplication of material in this work.

Contents

Contributors

Peter J Hoskin
Consultant Clinical Oncologist at
Mount Vernon Hospital,
Northwood,
Middlesex, UK
and
Professor of Clinical Oncology,
Royal Free and University College
London Medical School,
London, UK

Catherine Coyle
Consultant in Clinical Oncology,
Cookridge Hospital,
Leeds, UK

Dan V Ash
Consultant in Clinical Oncology,
Cookridge Hospital,
Leeds, UK

Anthony Flynn
Principal Physicist,
Cookridge Hospital,
Leeds, UK

Carolyn Richardson
Senior Physicist,
Cookridge Hospital,
Leeds, UK

Peter Bownes
Principal Physicist,
Mount Vernon Hospital,
Northwood, Middlesex, UK

Introduction

Peter Hoskin, Catherine Coyle

Brachytherapy is the delivery of radiation therapy using sealed sources which are placed as close as possible to the site to be treated. The very term 'brachytherapy' means 'near treatment'. It is therefore applicable for the treatment of tumours where a radiation source can be placed within a body cavity, for example the uterus, vagina, oesophagus, or bronchus or where the tumour is accessible to needle or catheter sources being placed within it, for example the breast, head and neck, prostate, and skin. In fact, brachytherapy has potential applications to most tumour sites. It can be used as primary treatment or in combination with external beam radiotherapy.

The principal advantages of brachytherapy lie in the physics of the dose distribution around a radiation source, which results in a high concentration of dose immediately around the source, and a rapid fall-off of dose away from the source with distance according to the inverse square law. Other advantages include the consequence of accurate localization of the gross tumour volume and immobilization of the area to be treated unlike the situation with external beam treatment when organ movement and set up errors are introduced. The major disadvantages of brachytherapy lie in the need to access the tumour, often with an operative procedure and the requirement for skilled personnel to undertake the treatments.

In the early days of radiation therapy brachytherapy using radium sources was a novel means of delivering a radiation dose to a tumour. The development of high energy external beam machines and the recognition of the problems associated with radiation protection in the use of radium resulted in a move towards caesium sources and the development of the concepts of afterloading in brachytherapy. Afterloading refers to the use of non-radioactive templates, tubes, or needles to define the implant which are then later loaded with the active radioisotope. Early systems used manual loading, typically using iridium wire which is still used particularly for head and neck techniques using hairpins or loops. In the modern era, high dose-rate remote afterloading machines using high activity small iridium sources are in wide use for a range of applications.

Sources may be regarded as continuous sources, that is with even distribution of the isotope along their length often encompassed within a sheath, for example, iridium wire, point sources or seeds, and discontinuous sources, or source trains which may be uniform or non-uniform in their dose distribution along their length, such as that found in remote afterloading caesium machines.

Dose-rate is an important consideration in the radiobiology of brachytherapy. Three dose-rate bands are now defined; low dose-rate based on radium (<1 Gy per hour), medium dose-rate, for example caesium (>1 to <12 Gy per hour) and high dose-rate (>12 Gy per hour). When changing dose-rate for a particular technique, it is important to remember that increasing the dose-rate increases biological effect and therefore, demands dose reduction.

There have been significant developments in external beam radiotherapy techniques over the last decade with the widespread introduction of multi-leaf collimation and conformal three dimensional planning techniques. Intensity modulated radiotherapy is likely to expand and become a standard technique in major cancer centres worldwide enabling irregular shaped volumes and dose variations across a volume to be routinely achieved with standard equipment and operating software.

This represents a major challenge to brachytherapy which used alone offers the optimal approach to conformal radiation therapy delivery. In combination with external beam radiotherapy, brachytherapy enables dose escalation within an intensity modulated radiation therapy volume with superior dose volume constraints to surrounding organs at risk compared to external beam techniques. It is therefore essential that expertise in these techniques is maintained and developed.

The following chapters seek to provide the reader with a sound infrastructure in the physics and dosimetry of brachytherapy followed by practical guides on the use of brachytherapy in common disease sites. Whilst low dose-rate, medium dose-rate and high dose-rate techniques are covered, more emphasis is put on high dose-rate afterloading techniques which are likely to replace most other forms of brachytherapy over the next decade. The advantage of these machines is that despite the requirement for fractionated treatments they allow short radiation exposure times and replace several days of inpatient treatment with outpatient therapy. The small source size <2 mm in diameter within flexible catheters which can be delivered to many body sites results in a very versatile treatment technique and the use of varying dwell times throughout the active length of the applicator means that each treatment can be individualized to ensure optimal coverage of a defined planning target volume. The radiobiology of high dose-rate introduces the concept of fractionation into brachytherapy but with increasing use and familiarity of standard schedules the initial uncertainty which surrounded this approach is rapidly fading.

This book has been written to provide a practical guide to the use of brachytherapy in current practice. The contents are not exhaustive and it is recognized that there will be many individual variations in practice, brachytherapy remaining one of the few arts in oncology which allows the practitioner to optimize treatment through individualization and experience. For this reason suggested further reading is given at the end of each section for those readers wishing to gain more detailed background information.

The authors are current practitioners at Mount Vernon Hospital, Middlesex (PJH) and Cookridge Hospital, Leeds (CC) in the UK. They organize the annual joint brachytherapy teaching course conducted through the Royal College of Radiologists in the UK on which the contents of this book are based.

Chapter 1

Isotopes and delivery systems for brachytherapy

Anthony Flynn

1.1 Introduction

The aim of this chapter is to consider which radioactive materials might be suitable as brachytherapy sources, to describe some practical brachytherapy sources, and to consider how these sources may be applied to the patient in a controlled and safe way. It is assumed that the reader has some familiarity with the basics of radioactivity and the emissions from radioactive substances.

Radioactive sources for brachytherapy are now available with many radionuclides and in various shapes and sizes, and there is no such thing as an 'ideal' brachytherapy source. Different sources have different applications depending on their emission type and radiation energy and how they are constructed. However it is possible to state some basic requirements:

1 The radionuclide must have a half-life of a few days or more for permanent implants and preferably at least a few weeks for temporary implants. Large corrections for radioactive decay during a temporary implant should be minimized. A long half-life for stock sources will give them a useful working life. Very short half-lives (e.g. a few hours) are unsuitable.

2 The energy of the emitted radiation should be sufficient to treat the required application, but not so high that radiation protection becomes difficult. Charged particle emission should be absent or effectively screened (except for beta emitters). Most modern brachytherapy sources emit photons of between about 0.35 to 0.66 MeV, though there are low-energy exceptions to this.

3 The radioactive material should be in a physical form that is insoluble and non-dispersible and can normally be encapsulated into a structure to prevent dispersion of radioactive material.

4 The radioactive decay process should have no gaseous or liquid decay products.

5 The material should be available in a high specific activity.

6 The sources should be available at reasonable cost so that the treatment does not become financially prohibitive.

Most brachytherapy sources are termed 'sealed sources'. These are constructed so that the radioactive material is encapsulated (often doubly encapsulated) in a container to minimize the risk of loss of radioactive material. Some sources, for example iridium wires, are not so encapsulated but nevertheless the risk of dispersal is very small; these are termed 'solid sources'. Sealed sources have to be wipe tested for leakage on a regular basis [1]. Preparation equipment for solid sources has to be checked regularly for contamination, even though this hazard is unlikely in practice.

1.2 Production of radioactive materials

The radioactive materials used in brachytherapy sources are produced by either neutron activation or are a product of nuclear fission. In neutron activation, often called the 'n-γ' reaction, a sample of a stable isotope of the element is placed in a neutron field in a nuclear reactor. Some of the nuclei of the element capture a neutron and become a radioactive isotope of the element, with the emission of gamma radiation. An example is the production of the radioactive iridium-192 from the stable iridium-191:

$$^{191}_{77}\text{Ir} + \text{n} \rightarrow ^{192}_{77}\text{Ir} + \gamma$$

With this production method the product will contain a mixture of the stable isotope and the radioactive isotope. The activity of the product will depend on the neutron flux and energy, the probability of nuclei interacting with the neutrons, the length of time in the reactor, and the half-life of the product itself.

Some radioactive materials used in brachytherapy are fission products. The process of nuclear fission arises when large nuclei divide and produce new elements, which are radioactive. An example is caesium-137 which arises from the fission of uranium and therefore is a by-product of the fuel rods in a reactor. The desired product has to be separated from other fission products. A more detailed treatment of production methods and yield can be found in reference 2.

1.3 Radionuclides in brachytherapy

This section includes brief descriptions of some of the radionuclides used for brachytherapy and these are summarized in Table 1.1.

1.3.1 Radium-226

In the early days of brachytherapy radium-226 and its daughter product radon-222 were the only radioactive materials used. Radium-226 is part of the radioactive series starting with uranium-238 and ending with the stable isotope lead-210. It has a half-life of 1620 years and decays by alpha emission to radon-222.

Table 1.1 Physical characteristics of some radionuclides used in brachytherapy

Source	Usual form	Production	Half-life	Emissions
Radium-226	Tubes, Needles	Naturally occurring	1620 y	2.45 MeV (max) gamma (from daughters when encapsulated)
Caesium-137	Tubes, Needles, Afterloading	Fission Product	30.17 y	0.662 MeV gamma
Cobalt-60	Tubes, Afterloading	Neutron activation	5.26 y	1.17, 1.33 MeV gammas
Iridium-192	Wires, Afterloading	Neutron activation	74 d	0.38 MeV (mean) gamma
Iodine-125	Seeds	Daughter of Xenon-125	59.6 d	27.4, 31.4, 35.5 keV X-rays
Palladium-103	Seeds	Neutron activation	17 d	21 keV (mean) X-ray
Gold-198	Grains	Neutron activation	2.7 d	0.412 MeV gamma
Strontium-90	Plaques	Fission Product	28.7 y	2.27 MeV beta particles
Ruthenium-106	Plaques	Fission Product	1.02 y	3.54 MeV beta particles

$$^{226}_{88}\text{Ra} \rightarrow \, ^{222}_{86}\text{Rn} + \, ^{4}_{2}\text{He}(\alpha) + \gamma$$

The daughter product radon-222 is radioactive and some of the subsequent daughter products decay by beta and gamma emission. The net effect is that an encapsulated radium source emits a complex photon spectrum with a maximum energy of 2.45 MeV, the alpha and beta particles being absorbed in the encapsulation. The radium in the form of radium sulphate powder was doubly encapsulated, usually in platinum, into tubes and needles and was also used in short distance teletherapy units. It was used extensively for treatment of cancer of the uterine cervix and implants. Dosimetry systems such as the Manchester System (see Chapter 2) were developed for it. However, it has several disadvantages, including the high photon energy requiring thick shielding, the risk of damage to a tube with consequent ingestion of the radium salt, and the biological harm that could result from accidental ingestion of the alpha emitting nuclides. The clinical use of radium has been discontinued in the UK and most other countries, as other more convenient radionuclides are now available.

1.3.2 **Caesium-137**

Caesium-137 is a product of uranium-238 fission. It decays by beta-minus emission with a half-life of 30.17 years and emits a photon energy of 0.662 Mev.

$$^{137}_{55}Cs \rightarrow \, ^{137}_{56}Ba + \, ^{0}_{-1}e + \gamma$$

It can be incorporated into glass beads and made into a variety of radioactive sources by encapsulating in stainless steel. It became readily available in the 1960s and largely replaced radium during the mid 1970s. It was considered safer than radium as its lower photon energy eased the radiation protection requirements and its solid physical form and lack of alpha emission made it safer in the event of a tube being damaged. Figs. 1.1 and 1.2 show a typical caesium-137 tube used for manually inserted gynaecological applications and a spherical caesium-137 source as used in a low dose-rate afterloading machine respectively.

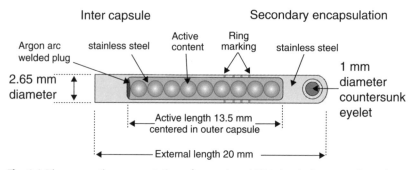

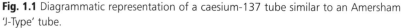

Fig. 1.1 Diagrammatic representation of a caesium-137 tube similar to an Amersham 'J-Type' tube.

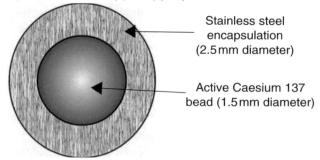

Fig. 1.2 Diagrammatic representation of an LDR afterloading caesium-137 pellet.

1.3.3 **Cobalt-60**

Cobalt-60 is produced by the neutron activation of the stable cobalt-59. It has a half-life of 5.26 years and decays by beta minus emission as shown below:

$$^{60}_{27}\text{Co} \rightarrow {}^{60}_{28}\text{Ni} + {}^{0}_{-1}\text{e} + \gamma$$

It emits gamma energies of 1.17 and 1.33 MeV. It has been used in tubes and needles, similar to radium and caesium tubes, but its relatively short half-life makes it inconvenient for this. Its main use in brachytherapy is in the form of pellets for high dose-rate afterloading machines. These are similar in appearance and size to the caesium pellets shown in Fig. 1.2.

1.3.4 **Iridium-192**

As mentioned in 1.2, iridium-192 is produced by the neutron activation of iridium-191. It has a half-life of 74 days and decays by beta minus emission:

$$^{192}_{77}\text{Ir} \rightarrow {}^{192}_{78}\text{Pt} + {}^{0}_{-1}\text{e} + \gamma$$

The photon emission is a complex spectrum with a weighted mean of about 0.38 MeV. When freshly produced, iridium-192 is contaminated by small amounts of the radioactive iridium-194, which arises from the neutron activation of the stable iridium-193; however this has a half-life of 17 hours and rapidly decays to insignificance.

Iridium-192 is often used in the form of a wire, used for low dose-rate (LDR) manual implants. The central radioactive core is an iridium/platinum alloy surrounded by a 0.1 mm thick platinum sheath. It is available in 0.3 mm or 0.6 mm overall diameter as wires or pins (Fig. 1.3). The thin wire is normally

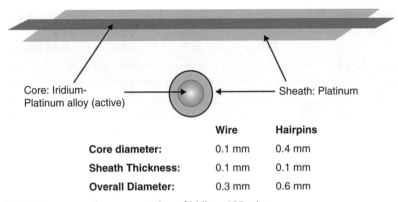

Core: Iridium-Platinum alloy (active)   Sheath: Platinum

	Wire	Hairpins
Core diameter:	0.1 mm	0.4 mm
Sheath Thickness:	0.1 mm	0.1 mm
Overall Diameter:	0.3 mm	0.6 mm

Fig. 1.3 Diagrammatic representation of iridium-192 wire.

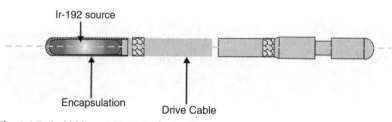

Ir-192 source

Encapsulation

Drive Cable

Fig. 1.4 Typical iridium-192 HDR afterloading source. Dimensions will depend on afterloader type.

supplied in coils of 500 mm length and is cut to the required lengths by the user. A hairpin has two 'legs' each of 60 mm length and a crosspiece that measures 12 mm. Their use is described in Section 1.6.2.2. Its high specific activity permits a high activity source to have small dimensions. It is therefore also used for high dose-rate afterloading machines (Fig 1.4), which are described in Section 1.6.3.2.

1.3.5 **Iodine-125**

Iodine-125 used principally for prostate brachytherapy (see Section 1.6.1), is a daughter product of xenon-125 which itself is produced by the neutron activation of xenon-124:

$$^{124}_{54}\text{Xe}\,(n,\gamma)\,^{125}_{54}\text{Xe} \rightarrow\,^{125}_{53}\text{I}\,+\,^{0}_{-1}\text{e}$$

Iodine-125 decays by electron capture and has a half-life of 59.6 days:

$$^{125}_{53}\text{I}\,+\,^{0}_{-1}\text{e} \rightarrow\,^{125}_{52}\text{Te}\,+\,\gamma$$

The gamma ray has an energy of 35.5 keV and characteristic radiation of 27.4 and 31.4 keV is also emitted. This low energy gives a half value thickness of lead of 0.025 mm, which makes radiation shielding highly effective.

For brachytherapy, iodine-125 is incorporated into implantable seeds. Several types of seeds are available from different manufacturers. As an example, Fig. 1.5 shows the structure of Oncura type 6702 and 6711 seeds. The 6702 seed has the iodine-125 absorbed on to an ion exchange resin but contains no radiographic marker. The 6711 seed has iodine-125 adsorbed on to a silver rod, which is encased in a titanium capsule. In this case the silver acts as a radiographic marker for imaging. The overall size of both seeds is 4.5 mm long and 0.8 mm diameter. The 6711 seed is the most commonly used type, particularly for prostate implants but the 6702 is sometimes used for temporary implants to other sites when a higher activity is desirable.

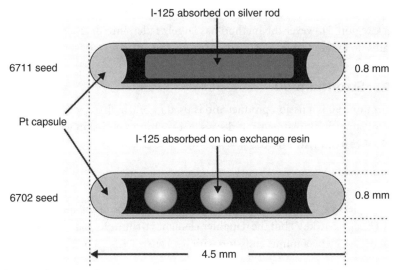

Fig. 1.5 Diagrammatic representation of two types of iodine-125. seed.

1.3.6 **Palladium-103**

Production of this isotope is possible via several reactions including the neutron activation of the stable palladium-102 and the interaction of protons or deuterons on rhenium-103. It decays by electron capture with a half-life of 17 days:

$$^{103}_{46}\text{Pd} + ^{0}_{-1}\text{e} \rightarrow ^{103}_{45}\text{Rh}$$

Like iodine-125 it emits a mixture of gamma rays and characteristic radiation, but with a slightly lower mean energy of about 21 keV. It is encapsulated into seeds of the same dimension as iodine-125 seeds and used in a similar manner for prostate implantation. Some operators prefer it for faster growing tumours citing a radiobiological advantage for the shorter half-life (i.e. higher dose-rate) but this benefit is not universally accepted [3].

1.3.7 **Gold-198**

This is produced by the neutron activation of the stable gold-197. It decays by beta emission with a half-life of 2.7 days to an isotope of mercury:

$$^{198}_{79}\text{Au} \rightarrow ^{198}_{80}\text{Hg} + ^{0}_{-1}\text{e} + \gamma$$

It emits a 0.412 MeV gamma photon (plus insignificant amounts of other energies). For many years gold-198 grains, consisting of gold-198 encapsulated in platinum, were used for permanent implants, especially for the head and neck region. However the method has largely fallen into disuse and gold-198 grains no longer feature in UK suppliers' catalogues.

1.3.8 Strontium-90

Strontium-90 is a fission product and is used in brachytherapy as a beta emitter for superficial treatments. It decays by beta minus emission with a half-life of 28.7 years to yttrium-90:

$$^{90}_{38}\text{Sr} \rightarrow {}^{90}_{39}\text{Y} + {}^{0}_{-1}\text{e}$$

The beta energy from strontium-90 itself is too low to be useful for brachytherapy ($E_{max} = 546$ keV) but the daughter element yttrium-90 (half-life 64 hours) also decays by beta minus emission with beta energy (E_{max}) of 2.27 MeV:

$$^{90}_{38}\text{Y} \rightarrow {}^{90}_{40}\text{Zr} + {}^{0}_{-1}\text{e}$$

Therefore the combination of strontium-90 with its daughter yttrium-90 provides a source of 2.27 MeV (E_{max}) beta particles with an effective half-life of 28.7 years. This material has been used for surface applicators, particularly ophthalmic applicators where the beta particles provide the required surface dose with rapid fall-off beneath. They were made from a strontium compound incorporated into a silver sheet, which is then formed into an applicator, with shielding on the reverse side. A variety of shapes, sizes, and concavities were available. Although no longer available from a major UK supplier, they are still in use in some hospitals. The surface dose-rate is such that the treatment takes a few minutes.

1.3.9 Ruthenium-106

This material has largely replaced strontium-90 in surface applicators, probably as it emits higher energy betas. It is a fission product and decays by beta emission to an isotope of rhenium with a half-life of 1.02 years:

$$^{106}_{44}\text{Ru} \rightarrow {}^{106}_{45}\text{Rh} + {}^{0}_{-1}\text{e}$$

The beta energy E_{max} is 3.54 MeV. The plaques are similar to the strontium-90 plaques in that the active material is incorporated into a silver sheet, which forms the surface of the plaque. The dose-rate is such that the treatment takes a few days, so the plaque is provided with suture holes to allow it to be fixed. Several shapes and sizes are currently available.

1.3.10 Other radionuclides

Various other radionuclides have been used or proposed for use in brachytherapy, including californium-252, phosphorus-32, samarium-145 and tantalum-182. These will not be discussed further as they are not in common use. More information is available in reference [2].

1.4 Definitions

The following terms are frequently used in brachytherapy, and it is useful to define them before describing the delivery systems:

Intracavitary Therapy. The insertion of brachytherapy applicators and radioactive sources into an existing body cavity, for example the uterine canal or vagina.

Interstitial Therapy. The insertion of brachytherapy applicators and radioactive sources directly into tissue, for example a needle or wires implant to the breast, floor of mouth etc.

Intralumenal Therapy. The insertion of brachytherapy applicators and radioactive sources into a lumen, for example the bronchus or oesophagus.

Intravascular Therapy. The insertion of brachytherapy applicators and radioactive sources into an artery. This treatment is used for the prevention of restenosis and may be applied to coronary or peripheral arteries, though the treatment techniques for the two are different. For reasons of space this will not be discussed further in this chapter.

Low, Medium and High Dose-Rate (LDR, MDR, HDR). There is no universally accepted definition of these dose-rate categories; the various bodies such as ICRU [4], AAPM [5] the UK Guidance Notes [1] suggest different boundaries between them. The reader is referred to reference [6] for further discussion of the various definitions. However, most users accept the following:

Low Dose-Rate (LDR). Dose-rates around 0.5 Gy.hr^{-1} to about 1 Gy.hr^{-1}. These are the dose-rates obtained or aimed for in traditional manual brachytherapy and it is generally accepted that dose-rate corrections to the prescribed dose are not required in this range (though there could be some debate about the upper end of this range). ICRU puts the upper limit at 2 Gy.hr^{-1} but most users would regard this as being MDR (see below).

Medium Dose-Rate (MDR). Dose-rates between about 1 Gy.hr^{-1} and 12 Gy.hr^{-1}. There is no sharp dividing line between LDR and MDR, but these are dose- rates where, although the therapy is still continuous, a dose correction for dose-rate effects is required. Many MDR afterloading cervix treatments are treated at about 1.5 to 2 Gy.hr^{-1}.

High Dose-Rate (HDR). Dose-rates greater than 12 Gy.hr^{-1} (0.2 Gy.min^{-1}). In practice most HDR machines operate at dose-rates much higher than this boundary, typically around 2 Gy.min^{-1}.

Pulsed Dose-Rate (PDR). This is a technique where high dose-rate 'pulses' of treatment (typically lasting five or ten minutes) are repeated at short intervals (typically once per hour). The intention is to simulate the radiobiological effects of LDR treatment using an HDR type machine. Clinicians who prefer the radiobiological effects of LDR can achieve this but with the flexibility of the complex dose distributions achievable by a modern HDR machine. The radiobiology is described in more detail by Brenner and Hall [7] and Fowler and Mount [8].

1.5 **Rationale for afterloading**

With minor exceptions, up to the late 1960s all brachytherapy sources and applicators were applied to or inserted into the patient manually, that is directly by the operator. Dr Robert Abbe at St Luke's Hospital in New York performed an early attempt at afterloading in 1905 but that was exceptional and manual placement remained the norm. As concerns grew about the possible effects of radiation on staff, there was increasing effort put into the reduction of radiation doses to operators and the techniques of afterloading were developed. Afterloading involves the initial placement of non-radioactive applicators or carriers in or on the patient followed by the subsequent insertion of the radioactive material.

- ◆ In 'manual afterloading' the sources are applied to the applicators by the operator using appropriate handling tools. This, of course, can only be done for low dose-rate treatments using low activity sources.

- ◆ In 'machine afterloading' or 'remote afterloading' a treatment machine applies the sources, usually under computer control. With appropriate shielded treatment rooms this technique permits high dose-rate treatments using high activity sources although the method is also used for low dose-rate treatments. Table 1.2 shows how various staff groups may benefit from the use of afterloading.

Table 1.2 Staff exposed to possible radiation hazard, showing the benefit obtained with afterloading

	Non afterloading	Manual afterloading	Remote afterloading
Physicists/Technicians	YES	YES	NO
Theatre staff	YES	NO	NO
Clinicians	YES	YES	NO
Radiographers	YES	PERHAPS	NO
Nursing staff	YES	YES	NO
Visitors	YES	YES	NO

1.6 Delivery systems

1.6.1 Manual application

Examples of manual application are the use of radium or caesium tubes for gynaecological treatments and the use of needles for implantation. These commonly use a dosimetry system, such as the Manchester System or the Paterson-Parker Rules and these are described in more detail in Chapter 2. Both of these have now been mainly superseded by afterloading methods.

Prostate implantation using iodine-125 or palladium-103 seeds is an example of a manual (i.e. non-afterloading) insertion method for a permanent implant. Early prostate implants used an open surgery retropubic approach but in the modern method the seeds are contained in needles, which are inserted through the perineum into the prostate. The position of the needles, the depth of insertion and the pattern of seeds in each needle allows a three dimensional arrangement of seeds to be deposited in the prostate. Typically about ninety seeds will be implanted using about twenty-five needles but the precise number depends on the size and shape of the prostate. The intended distribution of seeds in the prostate is determined from an ultrasound study of the prostate in which cross-sectional images 5 mm apart are acquired or, more recently, three dimensional prostate imaging is used. This may be done a few days in advance or, increasingly, at the same session as the implant. The needles are then loaded with the appropriate patterns of seeds, which are deposited in the prostate using ultrasound and fluoroscopic guidance. The low emission energy from the seeds and the ease with which it is shielded means that this technique does not have the same degree of radiation hazard as working with other radionuclides. Also the radiation hazard arising from the patient after implantation is minimal. An afterloading method is also available for prostate implants; both these techniques are discussed further in Chapter 7.

Beta emitting eye plaques (strontium-90, ruthenium-106) are also applied manually, but the radiation hazard is small provided care is taken to shield the active surface.

1.6.2 Manual afterloading

1.6.2.1 Intracavitary treatments

Systems for manual afterloading for cervix treatments were developed in the 1960s and are still current. Typically hollow carrier tubes (without the sources) are placed in the uterine canal and vagina in a configuration that is similar to the Manchester System and then their positions are checked by radiographs. Each source holder or train consists of low activity caesium-137 sources contained in a spring holder, which is attached to a labelled applicator handle. There will be a series of these source holders containing different numbers and activities of source. The source activities and their arrangement are chosen to match the activities required for the uterine and vaginal component of the (say) Manchester System. At commencement of treatment the source trains are withdrawn from the shielded storage container and inserted into the applicators; they are subsequently manually removed from the applicators at the end of treatment. Often the source holders are mechanically coded to the applicators to prevent incorrect loading of, for example, a train intended for the uterine canal into a vaginal position.

1.6.2.2 Interstitial treatments

Another common manual afterloading method is the use of iridium wire for interstitial treatments. Iridium wires themselves are described in Section 1.3.4. The dosimetry of these using the Paris System is discussed in Chapter 3 but here we will consider the implantation techniques. A pre-implant assessment or 'pre-plan' is usually carried out a week or two before the implant, at which point the approximate wires arrangement is defined and the wire activity calculated in order to give the required dose and dose-rate. The wire can then be ordered from the supplier for delivery a day or two before the scheduled implant date. The activity of the wire is checked before use.

The 'flexible tube technique' involves the insertion of plastic carrier tubing (the so-called 'outer tubing') through the site to be treated. The clinician implants these tubes through the tissue in the planned arrangement. The ends are loosely held by lead discs or clips, often with a nylon ball spacer to keep the clips away from the skin surface, but these are left loose at this stage. Often non-radioactive marker wires are placed in the tubing so that localization imaging can be done before the active wires are inserted. Meanwhile, the active wires are prepared. The appropriate lengths are cut from the coil and encapsulated into 'inner tubing' which is similar to but of smaller diameter than the outer tubing and a seal is placed in the tubing at each end of the wire to keep the latter in place. Commercial loading devices to assist with this process are available. As the wires are prepared they are placed in a shielded container and are individually identified, as in practice not all wires will be the same length. Following any imaging that may be required, the active wires are inserted into the outer tubing with forceps and using shielding when practicable. The clips are crimped to hold everything in place (Fig. 1.6). The expected removal time is determined from a Paris or other calculation based on a reconstruction of the imaging. When the wires are to be removed it is important that the tubing is cut between the nylon ball and the lead disc so that the active wires remain intact. If it is cut close to the skin there is a risk that the wire might be cut with a consequent risk of loss of radioactive material or its remaining in the patient. Also a radiation monitor must be used to check for complete removal of the sources from the patient. The operation of the monitor must be checked before removal of wires to ensure it is working. Untoward events have been reported by *Arnott et al.* [9].

One disadvantage of the flexible technique is that the wires are rarely straight and evenly spaced. This can be avoided by using templates and rigid needles instead of the flexible outer tubes. These are particularly suited to implants of the breast and perineum. For the breast a pair of templates can be used at each end of the needles to support them at the correct separation. In the perineum only one end of the implant is accessible, so a single thick template is used. Once the templates and needles are in place the prepared iridium wires are inserted into the needles, which are closed with screw caps. Localization imaging is often

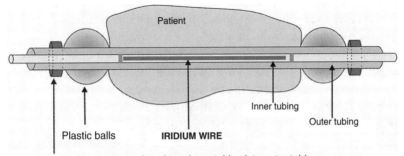

Fig. 1.6 Iridium wire encapsulation and fixing discs.

unnecessary for template implants as the implant geometry is known and fixed, and the dosimetry is predictable. The preferred removal method is to remove the needles and template intact so that the active wires can be recovered under shielded conditions in the preparation laboratory.

1.6.3 Machine (remote) afterloading

In this section afterloading machines from specific manufacturers will be referred to, for ease of description, but the reader should be aware that similar machines are manufactured by other suppliers. Reference to a particular machine does not imply endorsement of equipment from a particular supplier.

1.6.3.1 LDR/MDR afterloading

LDR and MDR afterloading will be discussed together as the equipment principles are the same, the only difference being in the source activities. One example is the Curietron (made by CIS), which was designed for the treatment of the uterine cervix, but is now almost obsolete. It used flexible preloaded trains of caesium-137 sources stored in a shielded safe.

The use of preloaded source pencils severely limits the use of this type of machine. The source configurations have to be decided at the time of purchase and cannot be changed until the sources are replaced. The Selectron (Nucletron), introduced in 1979 and still in use in many hospitals, overcame this difficulty by permitting source trains to be composed as required for each treatment. It is used predominantly for gynaecological treatments and it is available in both three and six channel versions. Each channel can be individually timed. The safe of a Selectron contains, in separate compartments, up to forty-eight caesium-137 sources and a large number of spacers. Sources and spacers are all spheres of 2.5 mm diameter. When the source loadings needed for a patient are known, the details are programmed into the machine and the trains are made up by pneumatically selecting sources and spacers in the correct order and loading them into channels in an intermediate safe. To start the treatment the source trains are driven pneumatically through transfer tubes into the applicators. The machine has fail-safe systems incorporated to ensure, for example, that the source trains cannot be driven out unless applicators are connected and to ensure that source position is correct within ±1 mm. The machine can be controlled from outside the patient's room to minimize radiation exposure to staff and, of course, can be interrupted for nursing procedures. During interruptions the source trains are returned to the intermediate safe. At the end of the treatment, the source trains are dismantled and sorted automatically, with the sources and spacers being returned to the main safe. LDR and MDR versions of the machine were identical except for the sources activity. LDR versions had a nominal source activity of 740 MBq (20 mCi) and gave about 0.8 Gy.hr^{-1} at the Manchester Point A and MDR versions had a nominal source activity of 1480 MBq (40 mCi) and gave about 1.6 Gy.hr^{-1} to the Manchester Point A.

The microSelectron-LDR (Nucletron) was introduced for interstitial therapy. This machine could remotely afterload up to eighteen iridium-192 ribbons (later strings of miniature caesium-137 sources were available). However, it suffered from the disadvantage that the iridium sources needed frequent replacement and the caesium version had the problem of restricted source combinations. It is now no longer available.

1.6.3.2 HDR afterloading

High dose-rate afterloading units came into use in the late 1960s. The advantage of high dose-rate treatments is the shorter treatment times (a few minutes rather than days), although the treatments have to be fractionated so overall treatment time is not necessarily shortened. However, particularly for cervix treatments, the short treatment times permit the use of rigid rectal retractors, which are likely to be more effective at reducing rectal dose. High dose-rate afterloading machines need to be housed in substantial shielded treatment rooms with appropriate interlocks, warning signs, patient monitoring equipment, etc. The first type to be installed in the UK was the Cathetron (made by TEM engineering which later became part of Varian), in Leeds, Charing Cross, Cardiff, and other hospitals in about 1968 [10,11].

As with the Curietron, the use of pre-loaded source pencils meant that source configurations were limited to those decided upon when the sources were installed, and were not likely to be changed for the five year working life of the sources.

A high dose-rate version of the Selectron (Nucletron) was introduced in the mid-1980s. The late 1980s saw the introduction of a new generation of afterloading machines using a single high activity iridium-192 source. The availability of small iridium-192 sources which contain typically an activity of 370 to 740 GBq (10 to 20 Ci) led to the development of HDR afterloading machines in which a single source is sequentially stepped through a series of dwell positions in all the treatment applicators in turn, thereby removing the need for several sources or source trains to be present in the machine. Such machines include the microSelectron-HDR (Nucletron) and its variant the microSelectron-PDR, the Gammamed and its PDR version (Isotopen-Technik Dr Sauerein and later MDS-Nordion), the Varisource (Varian) and others. These machines have a (usually) PC based control station outside the shielded treatment room. Here we will describe the microSelectron-HDR but the others operate on a broadly similar principle.

A single iridium source is contained in a tungsten-shielded safe in the head of the machine. The source capsule (Fig. 1.4) is laser welded to a long drive cable, which is connected to a stepper motor. The safe also contains a check cable assembly, which is identical in appearance to the source cable (but not radioactive) and is connected to a separate stepper motor. When the appropriate stepper motor rotates, the source or check cable is advanced into the treatment channel and can be positioned with an accuracy of ±1 mm. Within each treatment channel the source can occupy up to 48 dwell positions with a spacing of either 2.5 mm or 5 mm, thereby treating a maximum length of 235 mm. The exposure time at each of these dwell positions is variable and, in theory, could be different at each position. The front face of the machine, termed the 'indexer', has a series of numbered output ports to which the transfer tubes and applicators can be connected. The microSelectron has eighteen such ports but other machines may vary. This means that up to eighteen applicators can be connected simultaneously. When a treatment starts the check cable is driven in and out of the first channel to check connectivity and for obstructions. The source is then driven into the first channel and the programmed dwell positions and times are exposed. The source retracts back to the safe, the indexer advances one position (internally) and the check cable and then the source can now be driven out into the next channel, and so on until all the required channels and dwell positions have been exposed as programmed.

There are two principal advantages of these 'stepping source' machines over the older machines based on caesium or cobalt source trains. Firstly, the high specific activity of iridium-192 permits the source, and therefore the applicators, to have a small diameter. Typically the source has a diameter of 0.5 mm and the applicators have an external diameter of 2 mm. This means applicators are thin and flexible and can therefore be used in body sites not otherwise conveniently accessible to afterloading such as the bronchus and bile duct. Secondly, complex dose distributions can be produced from the large combination of dwell positions and times. Treatment planning systems are needed to plan these complex dose patterns; nowadays these can often be networked to the treatment unit so that the plan data can be transferred automatically, reducing the risk of transcription errors in entering the details into the treatment unit. Some machines combine the functions of treatment unit and planning system into one computer.

1.6.3.3 PDR afterloading

Afterloading machines designed for PDR use are physically similar to HDR machines. The main difference is in the operating software (and treatment

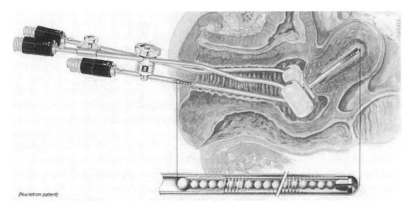

Fig. 1.7 Selectron applicators (cervix insertion), showing sources and spacers. Picture courtesy Nucletron.

planning software), which permits the programming of the treatment pulse duration, interval, and overall treatment time. Usually the iridium-192 source has a smaller activity than that used for HDR (typically 37 GBq) to avoid short dwell times which otherwise cannot be timed with sufficient accuracy.

1.7 Application of remote afterloading

In this section, we will briefly describe how remote afterloading may be applied in different ways to treat lesions in various clinical sites.

1.7.1 Intracavitary treatments

Most of the early afterloading machines were designed specifically around treatment of the uterine cervix. The applicator systems were based on the Manchester System [12] or the similar Fletcher System [13], consisting of an intrauterine tube and two vaginal source carriers (Fig. 1.7). Similar applicators were developed for the iridium-192 stepping source units, though these had a smaller diameter. For HDR machines the applicators often incorporated a rigid rectal retractor, which could be tolerated for the shorter treatment times and was considered to be more effective at reducing the rectal dose. Recently CT/MR compatible cervix applicators made from plastic materials have become available. Endometrial applicators are also available.

1.7.2 Interstitial treatments

The treatment of implants by remote afterloading was not feasible until miniature sources became available. The microSelectron-LDR, which uses iridium wires or caesium sources (described in Section 1.6.3.1), was one of the first such machines. The stepping source HDR machines made remote afterloading for interstitial treatments more attractive, provided that dose and fractionation modification required to change from LDR To HDR was accounted for. The small size of the source, dwell time flexibility, and large number of output channels makes them eminently suitable for multiple needle or flexible tube implants. The applicators are similar to the traditional manual treatments, that is flexible tubes or needles held in place by templates. Connectors link the needles or flexible tubes to the transfer tubes, enabling the source to be sent down, each in turn servicing the dwell positions as required by the treatment plan. The Paris System can be adhered to, but increasingly treatments are being modified by taking advantage of the dwell time and position flexibility of the afterloading source. Hospitals with appropriate operating theatre facilities can treat intraoperatively by which applicators are inserted into the tumour at operation and then subsequently connected to the afterloader.

1.7.3 Intralumenal treatments

Treatment of the oesophagus was possible with the Selectron-LDR [14], but other sites were not easily accessible due to the large diameter of the sources and applicator. However this is not the case with stepping source machines, and sites such as the bronchus [15], bile duct [16], and the larger diameter vascular system (such as the femoral artery) [17] could be treated. The applicators used for intralumenal applications are typically 2 mm in diameter and up to 1500 mm in length (Fig. 1.8). They can therefore be inserted through

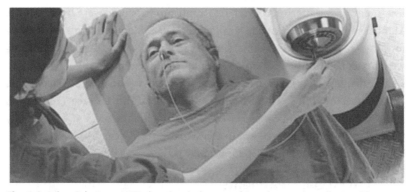

Fig. 1.8 MicroSelectron-HDR showing indexer and intralumenal applicator. Picture courtesy Nucletron.

the instrument channel of a suitable endoscope. The technique varies with site, but usually the treatment length and position is identified by taking radiographs with a marker in the applicator

1.8 **Surface moulds**

A surface mould is a custom-made device, which is attached to the patient in order to support an arrangement or radioactive sources at a known fixed distance, typically between 5 mm and 20 mm, from the surface to be treated. They are usually made to treat skin lesions but treatment of lesions at other sites such as intra-oral and intra-vaginal are also possible. Early moulds used radium-226 sources and later radon-222 seeds, gold-198 grains, caesium-137 tubes, and iridium-192 wires were used. Dosimetry systems such as the Paterson–Parker rules (see Chapter 2) allowed the calculation of the activity and arrangement of radium required to treat a given area to a prescribed dose. The dose to the treated area was often non-uniform due to the fact that the sources are discrete and the number of sources available was limited. A dose variation of ±10% of the prescribed dose was regarded as being acceptable. Surface moulds played a large part in early brachytherapy but they often contained substantial amounts of radioactive material and the radiation risk in their preparation, application, and removal from the patient led to a reduction in use. Most skin lesions were instead treated by superficial X-rays or electrons. However the advent of afterloading led to a reappraisal of surface moulds and many workers have used them through this method; this has been reviewed recently by Joslin and Flynn [18].

Later the method was adapted for the microSelectron, and various workers have treated surface moulds at HDR [20] and PDR [21].

References

1. Medical and Dental Guidance Notes. IPEM, York UK, 2002 (ISBN 1 9036313 09 4).
2. Aird E (2001). Sources in brachytherapy. In: Joslin C, Flynn A, Hall E (ed.). *Principles and practice of brachytherapy*. London: Arnold.
3. Pedley ID (2002). Transperineal interstitial permanent prostate brachytherapy for carcinoma of the prostate. *Surgical Oncology* 11:25–34.
4. ICRU Report 38, Dose and Volume Specification for Reporting Intracavitary Therapy in Gynaecology, International Commission on Radiation Units and Measurements, Bethesda, Maryland, 1985.
5. Remote Afterloading Technology (AAPM 41), published by American Institute of Physics for The American Association of Physicists in Medicine, New York, 1993.
6. Flynn A (2001). Afterloading systems. In: Joslin C, Flynn A, Hall E (ed.). *Principles and practice of brachytherapy*. London: Arnold.

7. Brenner DJ, Hall EJ (1991). Conditions for the equivalence of continuous to pulsed low dose rate brachytherapy. *International Journal of Radiation Oncology, Biology and Physics* **20**:181–90.

8. Fowler JF and Mount M (1992). Pulsed brachytherapy: the conditions for no significant loss of therapeutic ratio compared with traditional low dose rate brachytherapy. *International Journal of Radiation Oncology, Biology and Physics* **23**:661–9.

9. Arnott SJ, Law J, Ash D, Flynn A, Paine CH, Durrant KR, Barber CD, Dixon-Brown A (1985). Problems associated with iridium-192 implants. *Clinical Radiology* **36**:283–5.

10. O'Connell D, Joslin CA, Howard N, Ramsay NW, Liversage WE (1967). The treatment of uterine carcinoma using the Cathetron. *British Journal of Radiology* **40**:882–9.

11. Khoury GG, Bulman AS, Joslin CA (1991). Long term results of Cathetron high dose rate intracavitary radiotherapy in the treatment of carcinoma of the cervix. *British Journal of Radiology* **64**:1036–43.

12. Wilkinson JM, Moore CJ, Notley HM, Hunter RD (1983). The use of Selectron afterloading equipment to simulate and extend the Manchester System for intracavitary therapy of the cervix uteri. *British Journal of Radiology* **56**: 404–14.

13. Marbach JR, Stafford PM, Delclos L, Almond PR (1985). A dosimetric comparison of the manually loaded and Selectron remotely loaded Fletcher-Suit-Delclos utero-vaginal applicators. In: *Brachytherapy* 1984, The Netherlands: Nucletron BV.

14. Rowland CG, Pagliero KM (1985). Intracavitary irradiation in palliation of carcinoma of oesophagus and cardia. *Lancet* **326**:981–3.

15. Gollins SW, Burt PA, Barber PV, Stout R (1994). High dose rate intraluminal radiotherapy for carcinoma of the bronchus: outcome of treatment of 406 patients. *Radiotherapy and Oncology* **33**(1):31–40.

16. Montemaggi P, Costamagna G, Dobelbower RR, Cellini N, Morganti AG, Mutignani M, Perri V, Brizi G, Marano P (1995). Intraluminal brachytherapy in the treatment of pancreas and bile duct carcinoma. *International Journal of Radiation Oncology, Biology and Physics* **15**(2):437–43.

17. Waksman R, Laird JR, Jurkovitz CT, Lansky AJ, Gerrits F, Kosinski AS, Murrah N, Weintraub WS (2001). Intravascular radiation therapy after balloon angioplasty of narrowed femoropopliteal arteries to prevent restenosis: results of the PARIS feasibility clinical trial. *Journal of Vascular and International Radiology* **12**(8):915–21.

18. Joslin C, Flynn A (2001). Treatment of skin tumours. In: Joslin C, Flynn A, Hall E (ed.). *Principles and practice of brachytherapy*. London: Arnold.

19. Joslin CAF, Liversage WE, Ramsey NW (1969). High dose-rate treatment moulds by afterloading techniques. *British Journal of Radiology* **42**:108–12.

20. Svoboda VHJ, Kovarik J, Morris F(1995). High dose-rate microselectron moulds in the treatment of skin tumors. *International Journal of Radiation Oncology, Biology and Physics* **31**:967–72.

21. Harms W, Krempien R, Hensley FW, Berns C, Wannenmacher M, Fritz P (2001). Results of chest wall reirradiation using pulsed-dose-rate (PDR) brachytherapy molds for breast cancer local recurrences. *International Journal of Radiation Oncology, Biology and Physics* **49**(1):205–10.

Chapter 2

Principles of brachytherapy dosimetry

Carolyn Richardson

2.1 Introduction

The dose distributions are non-homogenous within the treated volume of a brachytherapy implant. Steep dose gradients occur and there is a sleeve of high dose surrounding each source. There are regions or plateaus of minimum dose between sources . Over the years several systems have been used to calculate and describe the dose distributions of brachytherapy implants, looking at the spacing and geometry of the source distribution and specifying how dose calculations should be made. The two systems to be discussed in this chapter are the Manchester system[1] and the Paris system[2].

Both systems used traditional dose formalism for manual calculation of doses and these were adapted for computer calculations. In 1995, the American Association of Physicists in Medicine (AAPM) published the report of Task Group 43 (TG43)[3]. The new formalism published in TG43 was originally used for iodine and palladium seed implants, but is now incorporated in many software packages for a variety of sources, including those used in afterloading treatment machines. More detailed consideration of the traditional and the TG43 formalisms will be found in the appendix at the end of this chapter.

It is important to remember that if any dosimetry system is to be used properly the sources must be distributed according to that system *and* the method of dose specification and prescription have to be adhered to.

How do we do these calculations in practice? Manual (non-computer) calculations are usually carried out with the help of tables or graphs of pre-calculated data. Examples would be Young and Batho tables for radium tubes, Cross-line graphs or escargot curves for iridium wire and Paterson–Parker tables for needle implants. Computer methods can be carried out using pre-calculated arrays of dose points around a particular type of tube or by dividing the source into a large number of point sources, and repeating point source calculations many times. Increasingly, computers are using the TG43 formalism for their calculations, and the rigid division between the Manchester and Paris systems has been eroded as computer algorithms use optimization techniques to derive the best dose distributions for any implant.

2.2 The Manchester system

The Manchester system was developed in the 1930s, and the system published in 1938[1], to improve on earlier techniques. It lays out standard treatments to be used, defines reference points at which the dose is to be specified, and defines source strengths to give a predictable and (almost) constant dose-rate to the reference point. The system was designed to improve the dose specification for a treatment and to be more scientifically rigorous than previous systems.

2.2.1 Manchester system for gynaecological brachytherapy

Gynaecological intracavitary treatments using radium–226 tubes were first carried out in the 1920s. Although the Manchester system was designed for use with radium tubes, it was easily adapted for use with caesium 137 tubes when these became widely available in the 1970s. For ease we will refer to the Manchester system as used with caesium from now on. The caesium tubes commonly available in the UK (J type, Amersham Health) are 20 mm long and 2.65 mm in diameter, with an active length of 13.5 mm.

The Manchester system applicators consist of a single intrauterine applicator and two vaginal source applicators (ovoids). There are three lengths of uterine tube, 20 mm, 40 mm, and 60 mm. These can be loaded with one, two, or three caesium J tubes respectively. The vaginal source applicators are hard rubber ovoids. These are all 30 mm long but come in three different diameters, small (20 mm), medium (25 mm) and large (30 mm). The uterine tube is inserted first and then the two ovoids are placed on either side of a spacer, and the whole insertion is packed tightly with radio-opaque packing to prevent movement. The system specified activity loadings for each uterine tube and ovoid. These were chosen such that they would give a dose to the defined reference point (point A) of 56.7 Rh^{-1} to 57.6 Rh^{-1} (equivalent to 54.5 cGyh^{-1} to 55.3 cGyh^{-1}). The only exception was if the short uterine tube was used, in which case the dose was approximately 12% less. This dosimetry system was calculated assuming that the applicators were perfectly inserted and totally symmetrical (Fig 2.1). The original definitions of points A and B were in centimetres and are referred to accordingly.

Point A was defined as being 2 cm lateral to the centre of the uterine canal and 2 cm superior to the mucous membrane of the lateral fornix along the line of the uterine canal. The dose at point A was thought to be representative

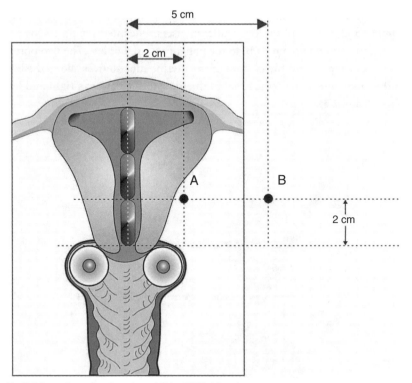

Fig 2.1 Location of points A and B in 1938 Manchester system.

of the minimum dose to most of the malignant tissue when treating cancer of the cervix. An additional calculation point, B, was defined to estimate the dose to the pelvic wall. Point B was placed 5 cm laterally to the patient's midline and at the same level as point A. When the system was first implemented, the calculations were all carried out for ideal insertions and no individual patient calculations were carried out.

The Manchester system was then developed as technology changed. Individual patient calculations could be carried out on radiographs. Radium was replaced by other nuclides, and afterloading systems were developed, both manual and later, computer driven. To accommodate these changes many users re-interpreted the definitions of the Manchester System. Point A was originally defined relative to the vaginal mucosa, but this is not visible on an X-ray film, therefore the definition was adapted so that the vaginal mucosa was assumed to be at the level of the cervix—practically, this meant the bottom end of the uterine radium/caesium tubes. When applying the calculation to individual patients, the insertion geometry is often not ideal, being neither symmetrical or on midline. This was resolved by amending the definitions of Points A and B in various ways by different users. One interpretation is to have two points A and two points B, to the patient's right and left. The two points A are 2 cm superior to the bottom of the uterine tube sources, along the line of the tubes, and 2 cm orthogonal to the tube, to the patient's right and left (Fig. 2.2 and 2.3).

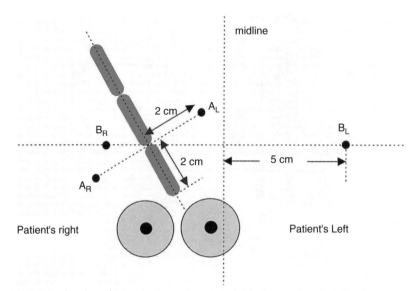

Fig 2.2 Application of Manchester system to individual patients—Anterior view.

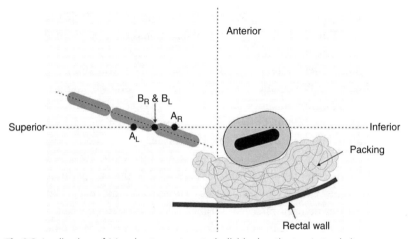

Fig 2.3 Application of Manchester system to individual patients—Lateral view.

The Manchester system can be used to calculate the prescription time easily, and can be adapted to afterloading machine techniques. To compare patients' treatments with those of another treatment centre, detailed information is required as to the definition of the reference points, and the doses received by them, the target volume, and the surrounding tissue.

2.2.2 Use of Manchester system for afterloading systems

In general, gynaecological afterloading systems have been designed with the Manchester system in mind. Designs such as the Fletcher applicator set incorporate the idea of a uterine tube with two ovoid applicators. The advantage of using such a set of applicators is that the applicators can be fixed in a rigid geometry, so that standard treatments can be given, according to the Manchester system, without problems of asymmetry associated with non-fixable tubes and ovoids.

The loading of the afterloader source trains, or planning of a stepper system source configuration can be arranged to mimic the Manchester source loadings. Thus the traditional clinical treatment can be given whilst utilizing the radiation protection advantages of afterloaders. The uterine tubes can still be used to define Points A and B, since the location of the tubes relative to the cervix is known (see Chapter 1).

2.2.3 Manchester interstitial system

The Manchester dosimetry system for interstitial implants was based on the use of radium sources[1]. It consisted of sets of dosage tables to calculate the amount of radium required and sets of distribution rules to determine how the radioactive material was to be distributed. The tables gave the product of the amount of radium (in mg) and the time (in hours) needed to give 1000 Roentgens to the treated surface.

The rules, known as Paterson–Parker Rules, were given for planar moulds, sandwich moulds, cylinder moulds, planar implants, and volume implants. The rules for planar implants are a simplified version of the mould rules.

Planar implants

The sources were implanted in a single plane and the dosimetry specified on a parallel plane, 5 mm from the sources plane. The implanted plane is divided into the periphery and the area. The sources are arranged as uniformly as possible on each of these categories, the proportion depending on the area. Distances between sources should not exceed 10 mm. The distribution rules for planar implants are laid out in Table 2.1.

A common arrangement for a planar implant is for a row of parallel needles with the ends 'crossed' by needles at right angles. If an end is 'uncrossed' 10% should be deducted from the area when reading from the table for each

Table 2.1 The distribution rules for planar implants

Area of implant	Peripheral fraction	Area fraction
<25 cm^2	2/3	1/3
25 to 100 cm^2	1/2	1/2
>100 cm^2	1/3	2/3

uncrossed end. For a two plane implant, planes should be parallel and the average area of the two planes is used for table reading purposes. The total activity should be divided pro-rata between the planes. The dose midway between the planes will be low by 10-30% depending on the separation and area.

Volume implants

For volume implants the implanted volume is divided into the 'rind' and the 'core'. The activity determined from the table is divided into 8 parts, and distributed as laid out in Table 2.2. Sources should be spaced as evenly as possible on each surface and within the volume, with not more than 10 to 15 mm between needles. A correction is made for 'elongation' when the volume dimensions are unequal. A correction is also made for uncrossed ends if necessary (−7.5% per uncrossed end).

2.3 **Paris system**

The Paris system[2] was developed for use in determining the dose distributions around iridium 192 wire implants. It is based on a set of implant rules and then specifies how dose calculations are to be made. This allows standardization of treatment and reporting. The Paris system can also be used to calculate doses for computer-driven afterloading systems such as iridium-192 high dose-rate afterloaders.

Table 2.2 The distribution rules for volume implants

Cylinder	middle part of the rind	core	each end
	4 parts	2 parts	1 part
Sphere	shell	core	
	6 parts	2 parts	
Cuboid	each side	core	each end
	1 part	2 parts	1 part

2.3.1 **Basic principles of the Paris system**

In the Paris system the active sources should be straight and parallel. Unlike the Manchester interstitial system, no crossing sources at the end of the implant are used. Ideally sources should be of equal length and of equal linear activity. They should be placed in a regular geometric pattern with equal separation between them. The separation may vary between 5 mm and 20 mm, depending on the number of sources, their activity, and the geometry of the implant. If the separation is less than 5 mm it is difficult to implant the sources in a precise regular manner, and hence the dose will be uneven; if the separation is over 20 mm then the high dose volume surrounding each source will be large, resulting in a greater risk of necrosis.

In cross-section the implants should have one of three geometrical arrangements. The simplest is a single plane implant, with the wires regularly spaced. In this case, a modification of the single plane implant can be a slightly curved plane, as for a chest wall, or a circular arrangement of needles/catheters such as that required for an anal margin treatment. To treat thicker tumours a multiple plane arrangement will be required. This can be a triangular pattern well suited to treating breast implants or for thicker tumours in the cheek or lip, a rectangular pattern, or square pattern. One particular application of the rectangular pattern is using hairpins to implant a tongue, where the legs of the hairpins define the two different planes.

2.3.2 **Paris system calculation**

The dosimetry is calculated on the central plane. The central plane is defined as a plane perpendicular to the sources midway along the sources. For an iridium wire implant, where the sources are not of equal length, the central plane should be placed as close to the mean midlength of the wires as possible, where the dose-rate between the wires due to their length contribution will be at its maximum. For hairpin implants, the central plane should be half way down the legs of the hairpin, ignoring the crosspiece, and orthogonal to the legs.

The Paris system dictates that the central plane must always be placed orthogonal to equal-length wires, which are evenly and regularly spaced. In practice, choosing the central plane requires a degree of experience. If the central plane is not orthogonal to the direction of the wires, the wires will appear to be further apart than they really are and the calculation will underestimate the dose-rate of the implant. The central plane should be perpendicular to the main direction of the source lines and pass through the estimated centre of the implant. Modern computer systems allow for rotation of the reconstructed implant in three dimensions to visualize easily the implant and choose the calculation plane(s).

Placement of the central plane for a large curved implant such as that for a floor-of-mouth treatment may be more subjective. The calculation should be discussed with the clinician so that the dose to the treatment volume required is calculated appropriately. It may be necessary to divide the implant into several sections and calculate a series of dose points accordingly.

Basal dose points

The basal points are defined on the central plane and are located at the points of minimum dose-rate between the wires. The arithmetic mean of all the basal dose-rates is used to calculate the basal dose-rate for the implant as a whole. The basal points can be defined geometrically. For a single plane, the minimum dose will be midway between each pair of wires. For a triangular arrangement of wires, the basal dose-rates are calculated at the centre of gravity (centroid) of each triangle (the intersection of perpendicular bisectors of the sides of the triangles), and for a square geometrical arrangement, the basal dose-rates are calculated at the centre of each square (Fig. 2.4).

Reference dose-rate

The reference dose-rate of the implant is defined as 85% of the mean basal dose-rate. This value was chosen to give an acceptable compromise between a steep dose gradient, whilst giving a reasonable contour coverage of the volume required.

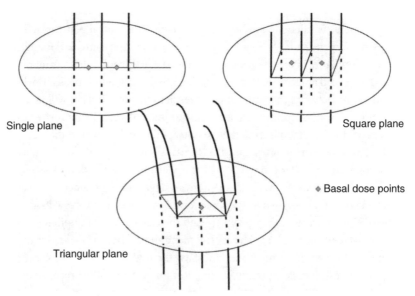

Fig 2.4 Geometry of Paris implants and basal dose points.

Hence, the treatment volume is defined as the volume enclosed by the 85% reference isodose.

When using the Paris system, only dose points on the central plane are required but modern planning computers can calculate dose points and iso-doses on multiple planes. Although the central plane is still used for defining the treatment time, it can be useful to carry out calculations on other planes to highlight hotspots or cold areas, with appropriate extra dose points, or isodose distributions, giving a fuller picture of the treatment given. This may be partic-ularly important when carrying out calculations from orthogonal radiograph reconstructions, since the soft tissues cannot be indicated in the calculations and so no dose volume histograms will be available.

Dimensions covered by the treated volume

The isodoses will 'pull in' between the wires, and hence it is necessary to allow extra coverage to give sufficient margins for the treatment. The length of the treated vol-ume is approximately 0.65 times the length of the source. This relationship is approximate, and depends partially on the wire separation, being relatively smaller for shorter wires. Therefore, for a particular target volume length, the sources should be about 20–30% longer, at each end, than the target volume (Fig. 2.5).

For a single plane implant the thickness of the treatment volume depends on the separation of the sources, and lies between 50 and 60% of that separa-tion depending on the number of wires and their length. For a two plane trian-gular implant, the thickness is about 1.2 times the source separation, and for a square arrangement the thickness is about 1.5 times the source separation. The lateral margins also vary depending upon the source separation and are best seen in a diagrammatical form. They are dependent on the source separa-tion and geometry (Figs 2.6, 2.7, and 2.8).

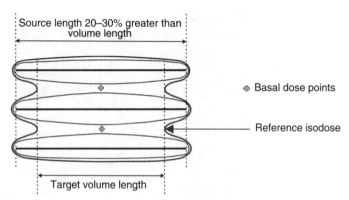

Fig 2.5 Length of treated volume of a Paris implant.

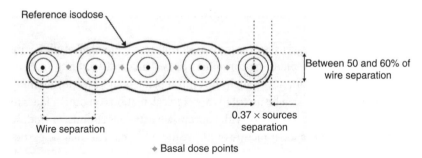

Fig 2.6 The dimensions of the treated volume margins for a single plane implant.

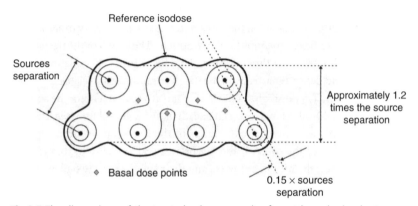

Fig 2.7 The dimensions of the treated volume margins for a triangular implant.

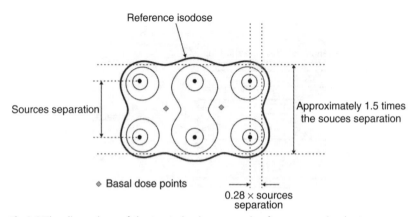

Fig 2.8 The dimensions of the treated volume margins for a square implant.

Calculation

If the wires are in a fixed geometry, such as in needles in a rigid plastic jig, then the calculation is simple and the Paris system can be easily applied. If the sources are located in a single plane on the patient's skin, then the separations can be measured directly, but usually the doses are calculated by reconstructing the implant from orthogonal films or CT scans. Care must be taken to correctly identify each wire. Once a geometrical reconstruction has been made, it can be used to provide the basis for a computer dosimetry calculation or a hand calculation based on dose-rate graphs, such as cross-line graphs. If a manual calculation is to be used, the dose-rate data is available in terms of unit strength of wire (AKR $1 \, \mu Gy \, h^{-1} mm^{-1} m^2$), and care must be taken to correct the calculation for the actual strength of wire used.

Modern computer systems may now take into account the decay of the iridium source whilst the implant is in place, and give an initial reference dose-rate. If not, it will be necessary to correct for the decay. The Paris literature has a correction table, which adds hours depending on the treatment time. An alternative is to carry out the calculation for the strength of the wire on the middle day of the implant.

2.3.3 Examples

Example 1

A single plane iridium wire implant consists of four wires (straight and parallel), each 50 mm long, and separation 15 mm. Wire strength (at mid implant), AKR = $450 \, nGy \, h^{-1} mm^{-1}$ at 1 m.

For a single plane implant, the basal dose points are placed midway between the wires and in this case we have three basal dose points between the four wires. The distance from each basal dose point to each of the four wires is measured. Using the suitable cross-line curve for a wire length of 50 mm, the dose-rate contribution from each wire to each point can be looked up, and thus the dose-rate at each point can be calculated. The cross line graphs give the dose-rates for wire of AKR $1 \, \mu Gy h^{-1} mm^{-1} m^2$. The dose-rates are then scaled for the actual strength of wire used.

For this example the dose-rates at the three points are 0.395, 0.417, and 0.395 $Gy h^{-1}$. Therefore the mean basal dose-rate is 0.402 $Gy h^{-1}$. To calculate the reference dose-rate the mean basal dose-rate is multiplied by 0.85, giving in this case a reference dose-rate of 0.342 $Gy h^{-1}$; therefore the time required to give a treatment of 65 Gy is 190 hours.

Table 2.3 Paris calculation example 2

	P1		P2		P3	
	Distance mm	Dose-rate Gyh^{-1}	Distance mm	Dose-rate Gyh^{-1}	Distance mm	Dose-rate Gyh^{-1}
1	11.55	0.24	11.55	0.24	23.1	0.095
2	23.1	0.095	11.55	0.24	11.55	0.24
3	11.55	0.24	23.1	0.095	30.5	0.062
4	11.55	0.24	11.55	0.24	11.55	0.24
5	30.5	0.062	23.1	0.095	11.55	0.24
Totals	0.877 Gyh^{-1}		0.910 Gyh^{-1}		0.877 Gyh^{-1}	

Example 2

Consider a two plane implant with five wires arranged in a triangular cross-section. Each wire is 70 mm long, and the separation is 20 mm. Calculate the air kerma rate of iridium wire required to give 25 Gy in 2.5 days.

Using trigonometry we can calculate the distances from each wire to each point. for example from P1 to Wire 1, 3 and 4, $10/x = \cos 30°$, therefore $x = 10/\cos 30° = 11.55$ mm; from P2 to wire 3 and 5, $20/y = \cos 30°$, therefore $20/\cos 30° = 23.1$ mm.

A table can then be constructed of the dose-rates for a wire strength of 1 μGyh^{-1}mm^{-1} at 1 m. See Table 2.3.

2.4 Comparison of Paris system and Manchester interstitial system

The Manchester system was designed to give the prescribed dose to ±10%, whereas the Paris system is based on the 85% reference isodose curve. For the Manchester system there are more sources with closer spacing between them, but in practice this means that the Paris system implant contains a larger high dose volume. For Manchester system, the sources are confined to the tumour volume; however, since the Paris system uses uncrossed wires, they need to extend beyond the tumour volume into normal tissue. A large drawback of the Manchester system is that the rules are very rigid. Any deviation from those rules will result in a bad dose distribution, whereas the Paris system is more flexible and can cope with calculating a poor implant. In both cases using a rigid template will guarantee a good implant.

2.5 **Optimization**

2.5.1 **Aim of optimization**

When treating with a stepping source afterloading machine, both the dwell positions and the dwell times can be adjusted to improve the dose distribution by keeping the prescription isodose as close as possible to the dose points specified and the doses as homogenous as possible throughout the implant. This allows the doses to normal tissue to be minimized.

There are a variety of computerized methods of optimization which are more fully covered in references (4) and (5). The dwell positions and times can also be set manually to allow minor adjustments to the computer since the mathematically optimized dose distribution may not always be the best clinically.

2.5.2 **Techniques of optimization**

Distance implants

For a distance implant, dose points are placed at a specified distance around the catheters. The optimization program calculates the dwell positions and relative dwell times so that the prescription isodose surface passes through these points. This type of calculation is used for single catheters, double catheters, and single plane implants.

Volume implants

A volume implant contains one or more planes of catheters. A series of dose points are placed inside the target volume midway between the catheters and throughout the implant. The relative dwell times are optimized to the same dose at those dose points.

If the dose point placement is complicated, the dwell positions can themselves act as dose points—this is known as geometric optimization and can be performed on distance or on volume implants.

2.5.3 **Practical use of optimisation**

Optimization can be used effectively for single line treatments such as for oesophageal tumours. If all the dwell times for a stepping source are equal for such a treatment, the overall longitudinal shape of the reference isodose will be similar to that of a cigar—broader in the centre of the treatment and narrower at the ends. If a cylindrical treatment is required, the isodoses can be made to conform far more to the cylindrical ideal by reducing the dwell position times in the centre with respect to the ends , and doses to organs at risk, such as the heart, can be kept to a minimum. For a single line treatment using regularly spaced dwell positions, the dwell times in the centre of an optimized treatment will be between a third and a half of the treatment times at the end positions.

2.5.4 Disadvantages of optimization

It is now possible to optimize the dose, say, to a breast implant, to ensure that the area is treated with as homogenous a dose as possible. Care must be taken however in altering treatments given in practice. The traditional Paris treatment with iridium wire will automatically result in the centre of the implant receiving more dose than the peripheral areas within the treatment volume. Changing to an optimized technique alters the dosimetry and no longer satisfies the Paris system. The active length of the catheters can be shorter, since the dwell times at the periphery of the volume treated can be increased to improve the coverage, by flattening the reference isodose surface at the outer ends of the catheters. The changes in dosimetry obtained by using optimization may be clinically desirable, but care must be taken in making such deviations from the clinically proven Paris system, and both clinicians and physicists should be clear as to how the optimization fully works before implementing any new technique.

2.6 Dose volume histograms

Dose volume histograms (DVH) play a useful role in evaluating the dose given to a treated volume by an implant. They are used in analysing what percentage of a volume receives what dose, and there are several variations in common use. For more detailed information see Chapter 5 in Reference 4. A DVH is represented as a graph with a series of dose intervals or bins (with a specified bin width) on the horizontal axis and a volume related to that dose interval, on the vertical axis.

A differential DVH plots the dose interval on the horizontal axis against the volume receiving that dose on the vertical axis. The plateaus of even dose between catheters in an implant will show up as peaks on the graph, and thus a differential DVH can be used to assess the homogeneity of dose distribution. A cumulative DVH also plots the dose interval on the horizontal axis, but the vertical axis plots the volume receiving *up to* and *including* that dose i.e. the volume encompassed by the relevant isodose. Cumulative DVHs can therefore be used to look for volumes of under- or over-dosing – cold or hot spots. When sampling a three-dimensional dose distribution, care must be taken to analyse a sufficiently large number of dose points for accurate results. When looking at individual sources, a large part of the variation in dose distribution is due to the inverse square law. Natural DVHs eliminate variations due to the inverse square law, giving the natural histogram of a point source to be a horizontal line. For these graphs, the narrower the peak, the more uniform the dose in the volume. When using any DVH to analyse implants, users should be clear as to the method used to calculate and present the information. Using DVHs is no substitute for examining the isodoses as superimposed on the required treatment volume, with the areas of normal tissue also marked.

2.7 **Reporting of brachytherapy treatments**

The International Commission on Radiation Units and Measurements (ICRU) has published two reports of particular use when reporting brachytherapy implants. ICRU Report 38 (1985) gives guidance on reporting absorbed doses and volumes in intracavitary therapy (6). It recommends that a combination of total reference air kerma, description of the reference volume and absorbed dose at reference points be used to specify intracavitary applications for cervix carcinoma (see p. 95). The original reference should be consulted for full details.

The main limitation of the ICRU 38 recommendations is that since its publication, new technologies have increased our ability to reconstruct implants. In particular, planning directly from CT scans gives a much better indication of doses both to the tumour and to normal tissue. Any new system of reporting will have to address the increased dosimetry information available and how it should best be presented to allow comparisons between different treatments.

ICRU Report 58 (1997) (7) was compiled to suggest methods of giving common agreement when describing interstitial implants and should be consulted for details of reporting such implants to allow intercomparison of implant techniques and clinical results.

2.8 **Conclusion**

Computer systems now allow image and source reconstruction using different imaging modalities and the temptation is to abandon the use of a particular system and use the computer calculation to see 'what is actually happening'. Any move away from traditional systems, however, must be done with great care, since it would mean treatments becoming more and more individual and hence clinical practice between centres would become impossible to compare and match. If an individual calculation is the only choice in an unusual case, care should be taken to fully document the treatment according to ICRU recommendations.

APPENDIX

Traditional dosimetry

The BIR/ IPSM published Recommendations for Brachytherapy Dosimetry as a report of a joint working party in 1993 (8). It recommends using absorbed dose to water in water specified in Gray. For a point source, the dose at a point P, r centimetres from a source of known Reference Air Kerma Rate (RAKR) is given by eqn (1).

$$\text{Dose} P = \text{RAKR} \times \frac{(\mu/\rho)_{\text{water}}}{(\mu/\rho)_{\text{air}}} \times \left(\frac{r_{\text{ref}}}{r}\right)^2 \times F \times t \tag{1}$$

where
RAKR is the reference air kerma rate;
$(\mu/\rho)_{\text{water}}/ (\mu/\rho)_{\text{air}}$ is the ratio of the mean mass energy absorption coefficients for water and air;
r_{ref} is the specified reference distance for the source measurement;
F is the combined attenuation and scatter factor;
t is the time of the exposure.

The ratio of the mean mass energy absorption coefficients for water and air is used to convert air kerma rate to absorbed dose in water. This ratio is constant at about 1.11 for photon energies emitted by clinically useful radionuclides (except I-125 where a ratio of 1.02 should be used).

With manual methods the absorption and scatter corrections are usually ignored, but they are taken into account in computer calculations. There are different ways of calculating F. The absorption and scatter of radiation in tissue will alter the dose distribution around interstitial and intracavitary radioactive sources. Within the treated volume this may be by as much as ±3% from that calculated in air, but at more distant points such as the pelvic wall in cervix intracavitary treatments the dose may be reduced by 10–15%.

One commonly used method is that published by Meisberger *et al.* in 1968 (9). Using experimental data they determined the ratio of exposure in water to that in air at distances up to 100 mm. Amalgamating this with experimental data from other investigators they fitted a third order polynomial that can be used for distances up to 100 mm from a source (eqn (2)).

$$F = A + Br + Cr^2 + Dr^3 \tag{2}$$

where r is the distance from the source and A, B, C and D are published coefficients for various radionuclides, which can be used with the equation to provide the scatter and attenuation factor. Other similar methods were developed, and then in 1992 Sakelliou et al. published eqn 3 (10), determined from Monte Carlo calculations.

$$F = 1 + ar + br^2 \tag{3}$$

Again, a and b are coefficients that depend on the photon energy, i.e. the radionuclide used and r is the distance from the source.

The traditional formalism works well for point sources, but real brachytherapy sources are not point sources. In accurate calculations it is necessary to correct for the physical shape of the source. The encapsulation material surrounding the source, such as stainless steel, will filter the rays emitted from the source. For a uniformly filtered line source, the thickness of the attenuation material will vary, depending on the relative position of the origin of the ray with respect to the point of calculation. To calculate the air kerma rate from a source a Sievert integral can be used (11). Some older computer software uses tabulated results of the Sievert integral.

The BIR/IPSM recommends an adaptation of the Sievert Integral, dividing the source into many small contiguous line source elements. Each element should be no more than 1 mm in length. Each line element is subjected to a different inverse square scaling, a different water absorption and scattering correction and to oblique filtration corrections for both the source material and the source encapsulation material (eqn (4)).

$$\frac{dD(r,\theta)_{\text{water}}}{dt} = \frac{(dK_{\text{air}} / dt)_{\text{ref}} 1.11 d_r^2 \sum \exp[-\mu_s t_s(\theta_i)] \exp[-\mu_a t_a(\theta_i)] f(r_i) / r_i^2}{N_x \exp[-\mu_s t_s(\pi / 2)] \exp[-\mu_a t_a(\pi / 2)]} \tag{4}$$

where $t_s(\theta_i)$ is the thickness of the encapsulation material at angle θ_i and $t_a(\theta_i)$ is the thickness of the source material at the same angle measured from the source axis. μ_s and μ_a are the corresponding filtration correction coefficients. The ratio of water absorption coefficient to air transfer coefficient is normally given the value 1.11 for most brachytherapy sources.

The traditional formalism used to determine two dimensional dose distributions in this way is only easily done for an isotropic point source. Clinical applications require a dose distribution in a scattering medium, and actual brachytherapy sources exhibit considerable anisotropy.

TG43

In 1995, the report of the American Association of Physicists in Medicine Radiation Therapy Committee Task Group 43 (TG43) was published (3). Their calculation uses measured or measurable dose distributions produced by a source in a water equivalent medium. It provides clear definitions of physical quantities, and all the equations required for the calculation of dose from a single source. The general formalism for the dose-rate, $D(r,\theta)$, for a two dimensional case, i.e. a cylindrically symmetric source, is defined according to eqn (5) for the geometry shown in Fig. 2.9.

$$D(r,\theta) = S_k \Lambda[G(r,\theta)/G(r_0,\theta_0)]g(r)F(r,\theta) \tag{5}$$

The symbols are defined in the following paragraphs. The parameters S_k and $G(r,\theta)$ are referred to a point on the transverse bisector of the source at a distance of 1 cm from its centre, i.e. $r_0=1$ cm and $\theta_0=\pi/2$ (units are in centimetres for consistency with original publication). The equation can be examined in five parts.

Air Kerma Strength (S_k)

The air kerma strength of the source, S_k is the source strength specified in air kerma rate at a specified distance along the transverse axis of the source. It is

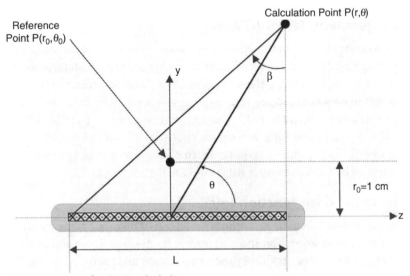

Fig 2.9 Geometry for TG43 calculation.

defined as the product of air kerma rate at a calibration distance, d, in free space, $K(d)$, measured along the transverse bisector of the source, and the square of the distance, d.

Mathematically this can be expressed by eqn (6).

$$S_k = K(d)d^2 \tag{6}$$

The unit of S_k is the 'U', where 1 U is 1 μGy m^{2}h^{-1}. It is customary to specify S_k in terms of a reference calibration distance, d_0, where the calibration distance must be large enough that the source can be treated as a mathematical point. This is usually chosen to be 1 m. In practice measurements are carried out in air and then corrected for attenuation. These measurements are normally carried out by the standardization laboratories (National Institute of standards and Technology, (NIST) in the US). The user then checks the source strength on delivery.

The dose-rate constant (Λ)

The dose-rate constant, Λ (units of cGyh^{-1}U^{-1}), is the dose-rate to water at 1 cm on the transverse axis of a source of 1 U in a water phantom. This includes the effect of source geometry, the spatial distribution of radioactivity within the source, encapsulation and self-filtration within the source and scattering by the medium (water). Λ is an absolute value and mathematically the dose-rate constant is $\Lambda = D(r_0,\theta_0)/S_k$. The numerical value of Λ also depends on the standardization measurements to which S_k is traceable.

The geometry factor ($G(r,\theta)$)

The geometry factor, $G(r,\theta)$, accounts for the variation of relative dose due only to the spatial distribution of radioactivity within the source. Mathematically this is an integration over the source core. For distances greater than about two to three times the largest source dimension it will differ from the inverse square law by less than 1%. For a point source approximation it can be reduced to $G(r,\theta) = 1/r^2$, and for a line source approximation it can be reduced to $G(r,\theta) = \beta/Lr \sin\theta$, where L is the active length of the source, and β is the angle subtended by the active source with respect to the point (r,θ).

The radial dose function ($g(r)$)

The radial dose function, $g(r)$, defines the fall-off of dose-rate along the transverse axis due to absorption and scattering in the medium. It can also be influenced by filtration of photons by the encapsulation and source materials. The radial dose function is defined only on the transverse axis (i.e. where $\theta = \pi/2$) and is unity at r = 1 cm.

The anisotropy function ($F(r,\theta)$)

The anisotropy function, $F(r,\theta)$, accounts for the anisotropy of the dose distribution around the source, including photon attenuation and scatter at any polar angle, θ, relative to that for $\theta = \pi/2$. It includes the effects of self-filtration, oblique filtration of primary photons through the encapsulating material, internal scattering within the source and attenuation and scattering in the surrounding water medium. There is a discrete value for each radial angle and distance.

References

1. Meredith WJ (1967). *Radium dosage: The Manchester system.* Edinburgh:Livingstone.

2. Pierquin B, Dutreix A, Paine CH, Chassagne D, Marinello G, Ash D (1978). The Paris system in intersititial radiation therapy. *Acta Radiologica Oncology* **17**:33–48.

3. Report of American Association of Physicists in Medicine Radiation Therapy Committee Task Group 43 (1995) *Medical Physics* **22**(2):209–35, Feb 1995.

4. Joslin CAF, FlynnA, Hall EJ (ed.). (2001). *Principles and Practice of Brachytherapy using afterloading systems.* London: Arnold.

5. Mould RF, Battermann JJ, Martinez AA, Speiser BL (1994). Brachytherapy from Radium to Optimization Ed., Nucletron International B.V, The Netherlands.

6. ICRU (1985) International Commission on Radiation Units and Measurements, Dose and Volume Specification for Reporting Intracavitary Therapy in Gynecology, ICRU Report 38 (International Commission on Radiation Units and Measurements, Bethesda, Maryland).

7. ICRU (1997) International Commission on Radiation Units and Measurements, Dose and Volume Specification for Reporting Interstitial Therapy, ICRU Report 58 (International Commission on Radiation Units and Measurements, Bethesda, Maryland).

8. Aird, EGA, Jones, CH, Joslin, CAF *et al.* (1993). Recommendations for Brachytherapy dosimetry. *British Institute of Radiology,* London.

9. Meisburger LL, Keller RJ, Shalek RJ (1968). The effective attenuation in water of the gamma rays of gold 198, iridium 192, cesium 137, radium 226 and cobalt 60. *Radiology* **90**:953–7.

10. Sakelliou L, Sakellariou K, Sarigiannis K, Angelopoulos A, Perris A, Zarris G, (1992). Dose rate distributions around ^{60}Co, ^{137}Cs, ^{198}Au, ^{241}Am, ^{125}I (models 6702 and 6711) brachytherapy sources and the nuclide ^{99}Tc^{m}. *Phys. Med. Biol* **37**(10):1859–1872.

11. Seivert R, (1921) Die Intensitätsverteilung der primären Gammastrahlung in der Nähe medizinischer Radiumpräparate. *Acta Radiol* **1**:89–128.

Chapter 3

Radiation protection issues in brachytherapy

Peter Bownes

3.1 Introduction

Everyone involved in work with radioactive substances should be trained in the principles of radiation safety. The three general principles of radiation protection that should be followed are:

- justification
- optimization
- limitation

Protective measures should always be in place to keep the dose levels as low as reasonably practicable (ALARP). In summary, a fundamental requirement is that the use of ionizing radiation should be optimized so that the patient receiving the treatment, and the hospital staff and the general public are not irradiated unnecessarily.

In this chapter, radiation protection issues will be discussed for sealed source brachytherapy only. It will address the regulations associated with brachytherapy and how they are implemented into safe practice in the clinical environment.

3.2 Radiation dose quantities and units

The term ionizing radiation is used to describe a variety of electromagnetic radiation and particle emissions. In brachytherapy the most important types are X-rays, gamma rays, alpha and beta particles, and neutrons. Ionizing radiation has the ability to separate electrons from atoms leaving the atom in a charged (ionized) state. Before the effects of radiation are discussed in a quantitative manner, it is necessary to introduce quantities and units associated with ionizing radiation.

3.2.1 Exposure

Exposure (X) measures the quantity of ionization produced in air by photons. The SI unit of exposure is coulomb per kilogram (C/kg). The original unit of exposure is the Roentgen (R) and is defined as 2.58×10^{-4} C/kg.

3.2.2 **Absorbed dose (*D*)**

Absorbed dose is the energy absorbed per unit mass in a material from any type of radiation. The SI unit of absorbed dose is the gray (Gy) where 1 Gy = 1 joule/kg. The original unit of absorbed dose is the rad, where 1 Gy equals 100 rad.

3.2.3 **Equivalent dose (*H*)**

Equivalent dose is the average absorbed dose in a tissue multiplied by a weighting factor that accounts for the biological effects of different types of radiation. This weighting factor depends on the type and energy of the radiation. For X-rays, gamma rays and beta particles this weighting factor equates to 1 and for neutrons and heavy particles (e.g. alpha particles) this weighting factor can take values between 5 and 20. Where there is more than one type of radiation the equivalent doses from each type are summed up. The SI unit for equivalent dose is the Sievert (Sv).

3.2.4 **Effective Dose (*E*)**

Effective dose incorporates a weighting factor for the sensitivity of the tissue irradiated. The weighting factors when multiplied by the equivalent dose to each tissue and summed for all tissues irradiated give rise to the effective dose. Table 3.1 shows the tissue weighting factors as defined by the ICRP[10] that reflect the radiosensitivty of each organ.

Table 3.1 Tissue weighting factors for effective dose calculations

Organ/tissue	Weighting factor
Gonads	0.2
Red Bone Marrow	0.12
Colon	0.12
Lung	0.12
Stomach	0.12
Bladder	0.05
Breast	0.05
Liver	0.05
Oesophagus	0.05
Thyroid	0.05
Skin	0.01
Bone surface	0.01
Remainder	0.05
Total	1

3.3 **Effects of radiation**

Ionizing radiation interacts with matter directly at the atomic level. DNA damage occurs through free radicals (hydroxyl radicals and hydrogen atoms) after ionization or excitation of water molecules. DNA is the critical target for ionizing radiation and an ionization event in a DNA molecule can cause a break (single or double) in a DNA strand. Factors that affect the radiation damage to cells include the type of radiation (how much ionisation is created along its path, the linear energy transfer (LET)), dose rate, total dose, dose fractionation, oxygenation status and stage of the cell cycle.

Data from survivors of Hiroshima, Nagasaki, and other epidemiological studies have informed the classification of biological effects. The effects of radiation can either be stochastic (random) or deterministic.

3.3.1 **Stochastic effects**

The probability (or risk) of an occurrence of a stochastic effect depends upon the total dose received and increases with dose received. The severity of effect is independent of the dose and dose rate. There is no threshold dose to which stochastic effects occur and similarly there is no threshold that exists below which safety is guaranteed.

These effects can be further classified into somatic and genetic effects. Somatic effects affect the individual who is irradiated whereas genetic effects affect subsequent generations as a result of the irradiation of reproductive organs. Examples of stochastic effects are cancer induction, where a group of mutated transformed cells proliferate in an uncontrolled manner, and genetic defects.

For stochastic effects it is assumed that any dose carries some form of risk. When setting radiation protection standards one must consider the risks associated with the radiation exposure and analyze the cost and benefits of the risk, related to the level of risk that society considers as acceptable.

3.3.2 **Deterministic effects (certain)**

There is a threshold dose for deterministic effects, above which deterministic effects are certain to occur. The severity of the effect may depend on the total dose received, the dose rate, dose fractionation and the sensitivity of the individual irradiated. Due to the existence of threshold levels for deterministic effects, permitted exposure levels can be set a certain safety margin below them. Examples of deterministic effects are skin erythema, acute radiation sickness and sterility.

3.4 **Current regulations and guidance**

The radiation employer is responsible for ensuring compliance with the regulations governing the use of ionizing radiation and for providing protective measures associated with their use. The radiation employer may be the NHS Trust, or a private company, a visiting contractor, self-employed person. The most important pieces of UK legislation governing the use of sealed radioactive sources in hospitals are described below. Similar regulations are in place in other countries.

3.4.1 **Ionizing Radiations Regulations 1999**

Each employer's radiation safety policy is based on their appropriate national regulations. The Ionizing Radiation Regulations 1999 (IRR99)[1] are the relevant UK regulations enforced by the Health and Safety Executive and are supported by the Approved Code of Practice[8] and the Medical and Dental Guidance Notes[9], which give practical advice to the implementation of the IRR99. IRR99 are concerned with the protection of all people at a place of work and the public at large.

The regulations require:

- A structured framework to ensure responsibilities are designated and understood.

- Adherence to the ALARP (as low as reasonable practicable) principle and commitment to optimization of procedures.

- Formal prior risk assessments on all equipment and procedures.

- Designation of controlled and supervised areas.

- Local rules and systems of work to ensure the restriction of exposure to staff.

- All staff to have appropriate training in radiation protection.

- Employers to have a radiation monitoring policy for staff and areas within their jurisdiction with adequate record keeping.

- A quality assurance program (preferably to a recognized standard) to exist, which will allow the continuous review and audit of all elements of a successful and safe brachytherapy department.

IRR99 also now covers outside workers (employees from another organization working in the hospital, e.g. service engineer), and includes regulations relating to pregnant staff.

Dose limits specified in IRR99, schedule 4 are detailed in Table 3.2 and are based on the ICRP 60 recommendations[10]. In addition to those in Table 3.2,

Table 3.2 Dose Limits per calendar year (schedule 4, IRR99)

	Employees (Aged 18 years or above)	Public
Effective dose (whole body)	20 mSv	1 mSv
Equivalent dose for the lens of the eye	150 mSv	15 mSv
Equivalent dose to skin (average 1 cm^2)	500 mSv	50 mSv
Equivalent dose to the hands, forearms, feet	500 mSv	50 mSv

for women of reproductive capacity (who are at work) the equivalent dose limit from ionizing radiation averaged throughout the abdomen is 13 mSv for any consecutive 3 month period. During a declared term of pregnancy the dose limit to the abdomen is 1 mSv.

Dose limits do not represent a dividing line between safety and danger, nor do they state what doses are acceptable. They should be put into the context of representing the upper level of risk that staff and public should be exposed to. By following the principle of ALARP it is intended to reduce the risks to levels far below those implied by the dose limits.

Classified workers are employees who are likely to receive an effective dose in excess of 6 mSv or an equivalent dose that exceeds 3/10 of any relevant dose limit. Most brachytherapy centres ensure, with compliance to local rules, that radiation doses received by staff are kept (ALARP) and in any case do not exceed 6 mSv in a year. It is therefore not necessary to designate any staff as classified workers.

3.4.2 Ionizing Radiation (Medical Exposure) Regulations 2000, (IR(ME)R)[2]

IR(ME)R regulations state the radiation protection framework required for justification and optimization of all medical exposures, to ensure adequate protection of patients and other individuals receiving a medical exposure. The roles and responsibilities are defined for the employer, referrer, practitioner, medical physics expert (MPE), and operator with guidance given for the duty holder. Schedule 1 of the regulations covers the necessary, written, employer's procedures to ensure safe clinical practice and a full audit system should be in place to review and update these when required. The final part of IR(ME)R covers the training required for practitioners and operators including theoretical knowledge and practical experience.

Table 3.3 Roles and responsibilities of duty holders in brachytherapy

Role	Main responsibilities
Employer	• Provide a radiation protection framework
	• Employer's written procedures for medical exposures and QA programmes
	• Identify duty holders and their local responsibilities
	• Ensure provision for adequate training
Referrer	• Must be a medical practitioner
	• Refers patient to IR(ME)R practitioner
	• Supply IR(ME)R practitioner with sufficient medical data
IR(ME)R Practitioner	• Must be an ARSAC holder for that procedure
	• Responsible for justification of a medical exposure, (implies all medical exposures involved in the brachytherapy treatment)
Operator	• Should be responsible for each practical step within the brachytherapy process
Medical Physics Expert (MPE)	• Closely involved in all brachytherapy practice
	• Provide advice on source and patient dosimetry, optimization and safety of treatment and treatment planning, on the QA programme
	• Consulted on new techniques and equipment (including acceptance and commissioning)

Roles and responsibilities identified for duty holders in brachytherapy are shown in Table 3.3. Decisions as to who can act as a referrer, IR(ME)R practitioner, operator, and MPE will be taken at a local level and documented in local protocols.

3.4.3 Radioactive Substances Act 1993[3]

The Radioactive Substance Act (RSA 93) controls the keeping and disposal of radioactive materials. It is intended to protect the general public against hazards, which could result from uncontrolled and unauthorized use or disposal of radioactive sources. Before being permitted to use, store or dispose off radioactive materials, the user must register with the Environment Agency. The certificates of registration and authorization will be specific to the user and the site. The certificate specifies the type of sources and nuclides, maximum number of each type of source and maximum activity of each registered source. Standard conditions are issued with the registration covering the supervision, marking, keeping, use, and records of the registered sources.

It also states what action to be taken if sources are lost or stolen and if a source is damaged. Before radioactive materials can be disposed, a certificate of authorization must be obtained. The justification of disposal and disposal routes must be clearly identified.

3.4.4 Medicines (Administration of Radioactive Substances) Regulations 1978[4] (Ammendment 1995)[5]

The IR(ME)R practitioner for brachytherapy must be authorized under the Medicines (Administration of Radioactive Substances) Regulations, i.e. the holder of an ARSAC certificate. Before being able to administer radioactive materials to a patient or volunteer, the clinician must apply for a certificate, which specifies each type of radioactive material to be used to the Administration of Radioactive Substances Advisory Committee, (ARSAC). The application should demonstrate that the clinician has had adequate training and experience and has available to them adequate scientific support.

3.4.5 Transportation regulations

All transportation of radioactive materials should conform to the current national and international regulations relating to the method of transport used. The competent authority in the UK for transport of dangerous goods is the Department for Transport. Before transportation of goods the radiation protection advisor must be consulted and a transport of dangerous goods safety advisor appointed.

Transportation of radioactive materials by road in the UK are covered by the Radioactive Material (Road Transport) Regulations 2002[6]. These regulations define: responsibilities of the consigner, consignee, carrier and driver; requirements for packaging; use of the correct labelling including UN numbers, shipping names, descriptions, transport index, and details of the consignee and consignor; consignment documentation for transport; appropriate training for drivers and others; placarding of vehicles; records of shipments; quality assurance programmes and audits of all aspects of the procedure.

If transfrontier shipments of radioactive materials are to be made between the European Union (EU) Member states, then EU regulation 93/1493/EURATOM must be followed. The consignee must obtain a EURATOM declaration from their national competent authority (Environment Agency in the UK) before shipment can take place, which declares to the supplier the consignee complies with all the relevant regulation.

Current regulations for shipment by air, rail, and sea will not be covered in this chapter but further information can be obtained from the relevant national competent authority. A useful website for the UK is //www.hmso.gov.uk.

3.4.6 **Health and Safety at Work Act 1974[7]**

The overall objective of this act is to secure the health, safety and welfare of persons at work and to protect the public from work activities. The act ensures control on the use and keeping of dangerous substances as well as control on the emission of noxious or offensive substances into the atmosphere. Outlined in the act are the general duties for employers, employees, and self-employed.

3.5 **Requirements for a brachytherapy centre**

When a hospital undertakes a brachytherapy service, it is essential that a radiation policy exists which describes how the use of ionizing radiations is safely and effectively managed within a legislative framework and employer's procedures. This will ensure that the radiation doses to staff, the public, and patients are kept as low as reasonably achievable.

The hospital policy should describe the responsibilities of the hospital, departmental heads, occupational medical advisors, RPAs, RPSs, MPEs and other staff involved with the use of ionizing radiation. To ensure that an effective radiation safety policy is implemented and maintained, it is vital that the employer establishes adequate communication and training for all areas of work within the organization.

A Radiation Protection Committee (RPC) should be formed and include senior management representatives, clinical and medical directors who use radiation, the health and safety manager, RPAs, and RPSs. The function of this committee is to review and oversee the implementation of advice and radiation protection arrangements. RPAs and RPSs should submit reports, at least annually, which include results of monitoring arrangements, radiation incidents, non-compliances, audits, and changes in practices.

3.5.1 **Responsible staff**

All employers who use ionizing radiation must appoint and consult a Radiation Protection Advisor (RPA). A RPA must be a trained scientist with recognized certification and appropriate knowledge who coordinates and monitors policy and practice across user departments. They should be available to give advice on health, safety, and legislation for ionizing radiation and should be consulted when any new process is considered or in the event of changes in practices or when incidents occur. The RPA's role also includes involvement with risk assessment, contingency plans, monitoring policies, and investigations of any radiation incidents.

To ensure compliance with IRR99 a Radiation Protection Supervisor (RPS) must be formally appointed, also an RPA. They should be full time employees of the section using ionizing radiation. Their role is to ensure that good working practice is maintained in order to keep radiation doses to staff and public as low as is reasonably practicable and ensure compliance to local rules and IRR99. Their main duties include acting as a first point of reference for practical radiation protection and maintaining staff awareness through training. They should assist with the derivation of local rules, contingency plans, and systems of work. The RPS should notify the RPA and the employer of any proposed changes in procedure and be involved in the introduction of any new techniques. Personnel dose monitoring, quality control and document control should also be in the remit of their job.

There is a legal requirement that an employer has a medical physics expert (MPE), who is a state registered clinical scientist and has at least six years appropriate experience. They should be closely involved in all brachytherapy practice and should work in close collaboration with professional colleagues. MPEs should provide advice on source and patient dosimetry, optimization and safety of treatment and treatment planning, the QA programme, and must be consulted on new techniques and equipment (including acceptance and commissioning).

3.5.2 Risk assessments

Formal radiation risk assessments must be carried out before any new activity involving work with ionizing radiation commences. Risk assessment should include:

- Location and description of the work
- Sources of radiation
- Identification of any hazards
- Indication of the personnel involved and who might be exposed.
- Evaluation of the risk in terms of severity and frequency.
- Dose monitoring and radiation surveys
- Control measures required to ensure ALARP, which may include appropriate personal protective equipment.
- Practicable steps required to prevent accidents and limit dose
- Actions to be taken

Radiation risk assessments should be made for all installations and areas where ionizing radiation is used. These assessments should be documented and reviewed when necessary, for example at a change in a procedure or workload. In addition to the radiation risk assessment general risk assessments for health and safety should be performed.

3.5.3 **Local rules**

Local rules describe the local arrangements for meeting the requirements of the appropriate regulations. They are in place to ensure restriction of exposure to staff and other persons and to minimize exposure in the event of a radiation accident. It is the responsibility of the employer to ensure that local rules are in place. Typically these are written by the local RPS with advice from the RPA and should be reviewed regularly. All employees are required to read them and sign a declaration stating that they will comply with them.

Local rules should include the following essential information:

♦ Identification and description of designated areas

♦ Names of responsible personnel, e.g. RPS, RPA

♦ Arrangements for restricting access to designated areas

♦ Dose investigation levels

♦ Summary of working instructions

♦ Contingency Arrangements

3.5.4 **Quality assurance programs**

An essential element to the radiation protection policy is to ensure that comprehensive maintenance and quality assurance programmes exist for all equipment. Quality control of all related documents is required and when changes are made they are passed on to all staff through relevant training. An equally important factor of the ALARP principle is the optimization of the required dose to the patient being treated. The potential treatment benefit will only be realized if a comprehensive quality assurance program exists to ensure that gross treatment errors are avoided, whether they are a result of equipment malfunction or human error.

3.6 **Designation of areas and systems of work**

Regulation 16 of IRR99 deals with the designation of controlled or supervised areas and it is the responsibility of the employer to identify them through assessment. A controlled area is designated if any person working in that area is likely to receive an effective dose greater than 6 mSv a year or 3/10 of any equivalent dose limits. Persons working or entering a controlled area must follow special systems of work to restrict exposure (ALARP), be trained and aware of emergency procedures, so they can prevent or limit the probability and severity of radiation accidents or effects.

A supervised area is designated if any person working in that area is likely to receive an effective dose greater than 1 mSv a year or 1/10 of any equivalent dose limits.

Radiation areas are designated as controlled areas to ensure that exposures from radiation sources are restricted appropriately. Areas adjacent to the controlled area should have exposures limited by the structural shielding around the controlled area. Shielding should reduce the levels of exposure to that required for a non-designated area. Where this is not practicable and the employer can exercise adequate supervision they may be designated as supervised areas. If areas are freely accessible to members of the public then the area must have exposures restricted to that of a non-designated public area.

Monitoring allows the determination of appropriate control measures for restricting exposures, to verify control systems are working effectively and to provide base estimates of personal dose. The effectiveness of shielding should be verified by radiation monitoring at the critical examination process to verify the structural design and the designation of areas. Instantaneous dose rate (IDR) measurements should be taken and time average dose rates (TADR), estimated over 8 hours, should also be calculated accounting conservatively for workload and usage. If occupancy factor is to be accounted for then the time average dose rate 2000 (TADR 2000), estimated over 2000 hours, must also be calculated. For guidance on the appropriate dose rates for area designations refer to the Medical and Dental Guidance Notes[9] Appendix 11. Designation of areas should be reviewed preferably annually by appropriate monitoring (IDR or cumulative dose using environmental monitors) or sooner if there is a change in the facilities or procedures that may effect the designation of areas or doses received by personnel.

Examples of permanently designated controlled areas are specially designed rooms for remote afterloading units and dedicated sealed source storage rooms. A theatre or recovery room in which radioactive sources are handled or housing a patient must be designated a temporary controlled area and this must be clearly indicated on all the doorways accessing the controlled area. Ward rooms/areas should also be assigned controlled area status when occupied by a patient containing radioactive sources. Specially designed rooms or lead shielding should be used as appropriate and dose rates should be measured at one metre to determine safe working practices and visitor restrictions. Temporary controlled areas can be returned to a non-controlled status after the radioactive sources have been removed. It is important that all nursing and theatre staff are properly trained.

3.6.1 Demarcation of controlled areas

Controlled and supervised areas should have their entrance demarcated with the appropriate warning sign at eye level, see Fig. 3.1. The sign should include information on the nature of the radiation source and the possible risks

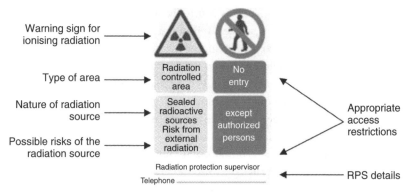

Fig. 3.1 An Example of an appropriate controlled area door sign.

associated with the source. There should be sufficient information to alert employees to the risks and to allow them to take appropriate action. It should also include restrictions on who is permitted to enter and when this is allowed (i.e. conditions of entry). For a remote afterloading brachytherapy unit, in addition to the warning notice there should be an illuminated warning light, which when illuminated indicates that the radiation source is in use (i.e. not in the protective safe position), this should be appropriately worded on the sign. All designated areas should be appropriately demarcated and should be described within the local rules.

3.6.2 Systems of work

Entry into controlled areas must be under a written system of work established to minimize exposures of persons working in them. The ALARP principle should form the basis of all systems of work. Three important radiation protection measures should be utilized as reasonably practicable:

- **Time** spent exposed to radiation should be minimized, as the total dose received is directly proportional to time.
- **Distance** from the source of radiation should be maximized, as the dose received follows the inverse square law.
- **Shielding** must be utilized between the source and personnel whenever practicable, which is achieved by placing a radiation absorbent barrier, often lead, between the source and the area to be protected.

Table 3.4 summarizes some general points that should be part of systems of work for theatres and wards. Personal dose monitoring is required to monitor doses to staff and ensure control measures are restricting the dose received to ALARP. All staff working in areas with ionizing radiation should be provided

Table 3.4 Examples of systems of work

Area	System of work procedure
Theatre	• Clear warning signs
	• Personal dose monitors and lead aprons should be used when appropriate
	• Only send for sources when required
	• Place sources behind protective barrier when not used
	• Check source information is correct
	• Use special tools to maximize distance when handling sources
	• Monitor equipment and area after use
	• Clear lines of communication between staff
Wards	• Clear warning signs
	• Personal dose monitors should be worn
	• Check number and position of removable sources against information provided
	• Define time limits and distance restrictions where practicable for staff and visitors
	• All operations should be carried out in the minimum time consistent with accuracy and safety
	• Maximize distance and utilize shielding, when appropriate
	• Monitor everything leaving the room to ensure no sources present. For permanent I-125 or Pd-103 implants of the prostate the urine is monitored for sources
	• Use special tools to maximize distance when handling sources
	• Shielded container available for a patient with removable sources
	• Monitor equipment and area after use
	• Clear lines of communication between staff

with whole body dosemeters and if applicable extremity monitoring. A system for collating the dose is required and dose investigation levels set.

3.7 **Contingencies**

All contingency procedures must be well documented and training courses for staff should be regularly scheduled. Staff should be aware of the seriousness of incidents in which a source may be lost or damaged and the protocols for dealing with these incidents should be clearly documented.

If a source is lost or stolen then the RPA and RPS must be informed immediately. The RPA has overall responsibility for ensuring that the source is located and returned to a secure shielded area, for investigating the incident and for informing the Health and Safety Executive and the Environment Agency.

Any suspected damage, breakage, or bending of a sealed source must be reported to the RPS or the responsible physicist immediately. The source must be tested for leakage immediately and then stored in a separate airtight shielded container. If the source is suspected to be leaking then the RPA should be informed immediately. All possible areas and personnel should be checked for contamination, and decontaminated if necessary. So far as reasonably practicable, the spread of contamination should be prevented. The RPA should inform the competent authority and arrange for the source to be returned to the manufacturer for disposal.

Each site should have readily available a current centralized sealed source inventory indicating the type, number of radioactive sources, and their location. This information along with a layout of the premises should be given to the local fire brigade.

Emergency procedures for remote afterloading units must cover the possibility of a source retraction failure. The risk of this occurring can be minimized by quality assurance of the unit and the applicators used. Figure 3.2 is an example of a high dose rate emergency procedure for a source retraction failure. It shows a hierarchic sequence of actions to be followed in the event a source fails to return to the safe position. A part of the daily QA procedure should ensure that all equipment required for the emergency procedure is available. All operators should attend a mandatory annual refresher course of the emergency procedure.

3.8 Quality assurance issues related to brachytherapy radiation protection

3.8.1 Source storage

All procedures related to receipt and storage of sources are designed to ensure compliance with RSA93[3]. A centralized sealed source inventory is required in order to ensure compliance to the Certificate of Registration issued by the Environment Agency and ensure that all sources are registered and no limits exceeded. The employer should appoint a custodian of radioactive sources to be responsible for the security during storage of the radioactive sources.

Radioactive sources should be stored in a safe containment, which is both shielded and secure. The safe should be compartmentalized, which allows easy

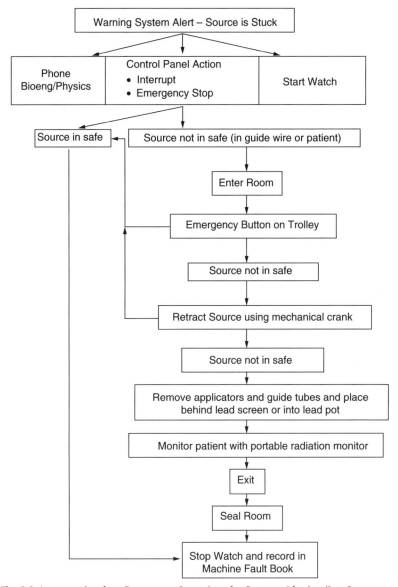

Fig. 3.2 An example of an Emergency Procedure for Remote Afterloading Source Retraction Failure.

access, control, and visualization of the individual sources. This type of storage system facilitates control of the sources and safe audits. Regular safe audits (stock checks) should be performed against the inventory to account for all the sources.

All sources should be identified individually with the following information recorded in the local inventory: type of radionuclide, physical and chemical composition, energy of emissions and decay scheme, air kerma strength on a specific calibration date, serial number or other distinguishing marks, date of receipt and the normal location of the source. Manufacturer's data sheets, source certificates, acceptance and leakage test reports should be stored with the local documentation. Documents should be kept for the period of time stated in the relevant regulations, the MDGN[9] summarizes the period documentation that should be kept, in Appendix 9.

The location of every source should be known at all times. The easiest method of achieving this is to have a log-book system, which should track the source at all times by adding an entry every time it is moved. Identifying responsibility for the sources, including handover of responsibility, and accountability of sources is essential.

Any radioactive sources, which have reached the end of their recommended working life, should be taken out of clinical use and be sent to a company authorized to dispose of that type of source or back to the manufacturer of the source. The return/disposal of sources should be documented in the inventory and the responsible physicist should follow the appropriate regulations and procedures taking advice from the RPA where appropriate.

3.8.2 Quality assurance checks on receipt of sources

All radioactive sources are delivered by the manufacturers with an accompanying calibration certificate of the **source strength**. The strength of all sources should be independently measured by the user and compared to the calibration certificate before clinical use. Measurements should be within ±5% of the calibration certificate and then for subsequent dose calculations the strength is taken from the calibration certificate, applying any required decay to the time of use. Records of source calibration and manufacturer's certificates should be kept.

Source strength may be measured using a re-entrant ionization chamber. For high activity sources, for example the 370 GBq high dose rate Ir-192 source in remote afterloaders, an ionization chamber can be used to measure the air kerma rate at a specified distance. Both methods require that the measuring device has a calibration traceable to a national standards laboratory and are described in more detail in Chapter 12.

Autoradiography is a simple method used to check the **distribution of activity** within a source. Autoradiography can also be used to check the configuration of pre-loaded source chains, and for remote afterloading machines, this technique can be used to check the precise location of the source positioning and the software that drives it. The **integrity of applicators** should be checked routinely by radiographic methods.

Sealed radioactive sources should be leakage tested by the manufacturer and the test certificate should be filed for future reference. If no leakage test certificate is provided then the user is required to test for both leakage and surface contamination before clinical use. Brachytherapy sources are required to be tested for leakage annually or when damage of the source is suspected. Leakage test consists of wiping the radioactive source using forceps with a swab moistened with water or methanol. The swab is then measured using a sodium iodide scintillation counter. If the activity of the measured swab is less than 200 Bq then the source is considered leak free. If the swab is measured greater than 200 Bq it may be a result of surface contamination from another source, therefore the source should be decontaminated and retested. If a leak is confirmed follow the contingency procedure in Section 3.7.

3.9 Source preparation (not including remote afterloading units)

Brachytherapy requests can only be made by an IR(ME)R practitioner with the relevant ARSAC certificate. When a formal request has been received, then the required sources can be clinically prepared. Transfer of sources should be accounted for at all stages of the treatment process along with the identification of the source custodian.

Radioactive sources should always be handled using long handled forceps and shielding should be utilized appropriately to reduce doses to the abdomen and protective observation screens used to minimize dose to the eyes. The speed of the operator is also important and when sources are not in use they should be stored in appropriate shielded containers. All handling and cutting tools, along with the area of preparation should be monitored for contamination at the end of source preparation. Manufacturers should always be consulted for instructions on sterilization of sources before clinical use.

Source transportation within the hospital should be done by trained staff following local procedures. The issue, distribution and return should be documented in the log book. Containers should be suitably shielded for the radioactive source being transported and be relatively easy to transport. Long-handled lead-lined transport pots are available which maximize the distance between the operator and the source and also have wheels for easy manoeuvrability. Transport containers should be labelled, packed appropriately, and secured. The transportation of sources should be done in the minimum time practicable and sources should never be left unattended. It is the responsibility of the person in charge of the transportation to ensure, by observation and verbal warning, that other persons do not enter the controlled area around the transport container.

3.10 **Treatment delivery and nursing care**

Most brachytherapy treatment rooms, used for manual loaded brachytherapy sources, have shielding included in the design, and additional mobile protective lead shielded barriers (generally 2 to 2.5 cm thick) can be used where practicable. Correct radiation warning signs should be displayed on the room entrance and a radiation detector, long-handled forceps, and a lead pot should be left in the room throughout the treatment. Long-handled forceps should be used to handle radioactive sources at all times, for example during the insertion and removal of sources for temporary implant treatments. If the patient is moved at any time then an appropriate warning sign should accompany them and be clearly visible. All sources must be accounted for at all times and documentation should illustrate this. For example with iodine-125 permanent implants of the prostate there is small risk that seeds may be passed in the patient's urine. Therefore to account for the seeds an X-ray film is taken at the end of the implant and then during the patient's stay the urine is monitored for the presence of seeds.

During the brachytherapy treatment the treatment room becomes a controlled area and staff must work to the local rules. For temporary implants, the time of insertion of the sources should be documented and displayed along with the proposed removal time. The custody of the sources is signed over to the nurse in charge of the patient who accepts responsibility for them. Restrictions on nursing time will be calculated for each patient and displayed on the warning signs. All nursing procedures should be carried out in the minimum time consistent with accuracy and safety of the procedure. Unnecessary time must not be spent in close contact with the patient. Visiting is discouraged during the treatment, but if necessary visiting time is kept to the restricted daily nursing time.

At the end of the patient's treatment temporary implant sources should be removed using long-handled forceps and placed in lockable shielded container. Care should be taken not to damage the sources during removal or cause any unnecessary trauma to the patient. The patient should then be monitored using a Geiger–Muller counter to confirm that all the sources are removed.

The sources should then be transferred back to the sealed source room and they should be checked for damage. The custody of the sources is transferred back to the source technician/physicist. The patient may leave when all the sources have been accounted for.

Sources and applicators post-treatment should be cleaned when necessary by being immersed in some bactericidal fluid before being stored. Manufacturers offer guidance on suitable ways to clean sources and thus prevent any chemical damage from cleaning agents. Sources and applicators once cleaned should be inspected for damage before being stored. All containers used in the process should be checked for contamination before storage.

3.11 **Patient related issues**

3.11.1 **Temporary implants**

All standard nursing procedures have been discussed in the previous section. Before patients can be discharged from the hospital, all the radioactive sources must have been removed and accounted for.

If surgical intervention is required following a therapeutic administration of a radioactive substance, while still under restrictions, then the RPS/RPA must be consulted to gain further information of the level of the hazard. Temporary implants should be removed from the patient before surgery commences. If a patient dies when a temporary implant is still in situ then the radioactive sources and applicators should be removed immediately and the body monitored before it can be released for a post-mortem examination or disposal. Appropriate instructions should be documented in the local rules.

3.11.2 **Permanent implants**

Information should be given to the patient about potential hazards and restrictions of contact between the patient and others. An instruction card should be carried at all times by the patient indicating the restrictions that apply and the period they apply, along with key information about the implant, radioactive source used, activity, and contact details of the hospital which performed the implant. Patients receiving permanent iodine-125 or palladium-103 seed implants of the prostate form a relatively low risk for the personnel involved and can be discharged on the day of the implant. They are advised to avoid prolonged periods of very close contact with children and pregnant women for the first two months post implant.

For patients who have received a permanent implant and who require surgery, the RPA/RPS should be consulted prior to surgery. For the example of I-125 prostate seed implants, there will be no restrictions on normal invasive medical care, but elective abdominal surgery should be avoided until the expiry of the instruction card. If delay is not feasible then RPA should give advice on a case-by-case basis, which should incorporate an individual radiation risk assessment.

If a patient dies after a permanent implant then as long as the activity is below the designated levels for that isotope[9], no special restrictions apply for burial of the corpse. For iodine -125 implants of the prostate the level of activity will not exceed the limit of 4000 MBq, therefore there are no restrictions required for burial. For cremation of a corpse generic radiation risk assessments and environmental impact assessment should be made to determine if cremation is permitted, which will depend on the time lapsed after administration and the total activity implanted. Cremation should be avoided during the period of restrictions but if unavoidable the sources must be removed at autopsy before cremation. Advice and control measures should be given to the employer at the crematorium to ensure doses are ALARP.

3.12 Principles of afterloading brachytherapy

Afterloading systems in brachytherapy have the major advantage of reducing radiation doses to staff and optimization of the dose distribution to the patient. Afterloading refers to any method where applicators are placed in the tissues or in the body cavities and the sources are loaded later. This allows time and care to be taken over the implant without the risk of receiving any radiation exposure. Applicator reconstruction either by standard radiographic checks (generally two orthogonal films) or sectional imaging can be done using dummy sources, which check the geometry of the implant and aid treatment planning.

Afterloading of sources can be achieved either manually, by hand, or remotely by specially designed micro-electronic equipment that drives the source or sources into position. Although manual afterloading techniques reduce the radiation dose to the clinician in theatre, the main disadvantage of this technique is that nursing staff are still exposed during nursing procedures and the radiographer/oncologist exposed during insertion and removal of the radioactive source.

In remote afterloading systems the radioactive source(s) are contained in a lead safe close to the treatment couch. The source(s) are then transferred into the required position by either a pneumatic transfer system (LDR/MDR Cs^{137} units) or via drive cables (Ir^{192} HDR units). The sources can be retracted from the patient and stored back in the safe if any nursing procedure is required during the treatment. Therefore, no radiation exposure should occur to staff unless the equipment malfunctions or interlocks fail.

3.13 **Brachytherapy room design**

Optimally brachytherapy treatments should be carried out in a specially designed shielded single room. However appropriate bed shielding could be added to a ward side room, additional structural requirements are required to be investigated if this room is not a ground floor room.

The RPA will be the project lead in room design and should ensure all the building contractors understand the importance of the shielding requirements. The RPA must ensure through all stages of the project, planning, installation, and commissioning stages, that the requirements of the design are correct.

There are many considerations and questions about the proposed use of the room, which need to be addressed at the initial stages of the design brief. The design of the brachytherapy facility must take into consideration the **intended use of the room** and any potential future use. This should include the proposed treatment techniques and the typical dose rates of these treatments, the number of patients to be treated, duration of the treatment, and whether there are additional uses for the room, which may need special requirements. Specific properties of the radionuclides to be used and the maximum air kerma strength used will determine the shielding requirements for the room. Table 3.5 shows some of the properties of the common nuclides used in brachytherapy.

The **location of the room** is very important, as it needs to be close to sealed source room, theatres, X-ray and other imaging modalities (CT and possibly MRI). The surrounding areas must be taken into consideration as they will impact in the shielding design of the room. All adjacent areas, including rooms above and below, must be surveyed at an early stage to ensure the shielding requirements are adequate.

The **size of the room and layout design** will depend on the type of treatments to be done. Adequate access of bed trolleys will need to be incorporated and thought should be given to the arrangements to be made for contingency plans and the most efficient and safe way of performing them. Additional considerations may include ensuite facilities or local bed shielding.

It is a requirement from IRR99 that radiation levels and associated hazards are estimated as part of the design stage, through prior risk assessment. Adequate design should then be confirmed through radiation monitoring. The dose limits were shown earlier in Table 3.2. Occupancy factors for the various surrounding areas can be incorporated into the dose estimates, NCRP Report 49[11]. For example a permanent work area has an occupancy factor of 1, while a waiting room, corridor, toilet and staircase have an occupancy factor

Table 3.5 Basic properties of common brachytherapy radioactive sources

Nuclide	Principal γ energy (MeV)	Half life	Typical reference Air kerma rate (μGy h⁻¹@1m)	Transmission through lead (mm) HVT	TVT	Transmission through concrete (mm) TVT
^{60}Co	1.17, 1.33	5.27 years	HDR remote afterloader up to 5.7×10^3	12	40	206
^{137}Cs	0.66	30.17 years	LDR remote afterloader 28 to 115	6.5	21	157
^{192}Ir	0.3–0.6	73.83 days	Hairpin/wire 0.13-1.3 per mm HDR remote afterloader up to 42×10^3	4.5	15	113
^{125}I	0.035	59.4 days	Seeds for permanent prostate implant 0.1 to 1 per seed	0.03	0.1	–

HVT = half value thickness

TVT = tenth value thickness

of 1/16. Brachytherapy sources are not collimated so the usage factor in all calculations will be 1. It is important to work to lower levels as the workload of the room may increase, techniques may evolve or current dose limits may be reduced. Whenever possible the dose rate outside the room should be reduced to a level at which it can be classed as a non-designated public area (Ref. 9, appendix 11). The treatment room will be designated as a controlled area and it will be necessary for local rules to specify systems of work for that area in order to restrict exposure.

Other requirements to be considered are:

- Doors must be lead lined (~6 mm of lead).
- The bed should not be in the direct line of the door.
- Staff working within the controlled area must be monitored.
- A portable Geiger–Muller monitor should be available near to, but outside, the treatment area.

- Radiation warning notice should be clearly visible at the entrances to the room.
- Suitable source container(s) and special handling tools should be available.
- CCTV and intercom system should be incorporated.
- For remote afterloading systems additional requirements are required:
  - For HDR remote afterloading units a small maze may be necessary to ensure no direct irradiation of the lead-lined door and to reduce scatter at the door.
  - Door interlocks should be present.
  - In addition to in-built radiation monitors of the system there should be an independent radiation monitor that indicates when the source is out.
  - An illuminated warning sign should indicate when the source is in the treatment position.
  - Emergency stop buttons to be clearly labelled and visible.
  - Emergency mechanisms should be in place for source retraction failure or power failure.
  - If a diagnostic X-ray unit will be used within the treatment room to obtain localization radiographs, then warning systems must be in place for this use and protection for the operator provided.

The IPSM report 75[12] shows examples of typical treatment rooms for a manual Cs-137 afterloading room for gynaecological intracavitary treatments, low dose rate/medium dose rate remote afterloading and high dose rate remote afterloading facility.

References

1. Ionising Radiations Regulations 1999, Statutory Instrument 1999 No. 3232. London: HMSO.
2. Ionising Radiation (Medical Exposure) Regulations 2000, Statutory Instrument 2000 No. 1059. London: HMSO.
3. Radioactive Substances Act 1993 (Chapter 12). London: HMSO.
4. Medicines (Administration of Radioactive Substances) Regulations 1978, Statutory Instrument 1978 No. 1006. London: HMSO.
5. Medicines (Administration of Radioactive Substances) Ammendment Regulations 1995, Statutory Instrument 1995 No. 2147. London: HMSO.
6. The Radioactive Material (Road Transport) (Great Britain) Regulations 2002, Instrument 2002 No. 1093. London: HMSO.
7. Health and Safety at work etc. Act 1974, London: HMSO.
8. Work with ionizing radiation. Approved code of practice and practical guidance on Ionizing Radiations Regulations 1999, London, HSE (2000).

9. Medical and Dental Guidance Notes. A good practice guide to implement ionizing radiation protection legislation in the clinical environment, Institute of Physics and Engineering in Medicine 2001.

10. ICRP Publication 60, International Commission on Radiological Protection. Radiation Protection–1990. Recommendations of the International Commission on Radiological Protection. Pergamon Press 1991.

11. NCRP Report 49, Structural Shielding Design and Evaluation for Medical Use of X-rays and Gamma Rays of Energies up to 10 MeV, 1976, NCRP Publications.

12. IPSM Report 75, The Design of Radiotherapy Treatment Rooms, IPSM 1997 (reprinted by IPEM 2002).

Chapter 4

The role of brachytherapy in head and neck cancer

Catherine Coyle

4.1 Introduction

Brachytherapy in head and neck cancer is languishing paradoxically at a time when interest in conformal radiation treatment is at its peak. As interstitial brachytherapy allows increased dose intensity and reduction in dose to surrounding organs at risk, it could be considered the ideal conformal treatment.

Its disadvantages lie in the lack of expertise, competition with surgical techniques, and radiation protection issues. It still has a place in several key clinical areas:

(1) primary treatment of small T1 and T2 squamous cell cancers

(2) in combination with external beam radiotherapy

(3) re-irradiation in previously treated areas either for recurrence or second primaries.

Expertise is concentrated in specialist centres with the majority using low dose-rate iridium implanation under the Paris rules. Similar techniques have been adapted to allow the use of high dose-rate afterloading systems, but this chapter will concentrate on the low dose-rate techniques. The most common indications are for the treatment of small tongue cancers, floor of mouth, buccal mucosa, and lip tumours. Careful assessment in the clinic to determine suitability for implantation is essential; a good knowledge of the Paris rules is necessary to pre-plan the implant.

(1) The tumour volume must be accurately demarcated—so patients with indistinct margins, surrounding dysplasia or erythroplasia are not suitable.

(2) The tumour site must be accessible.

(3) The site and surrounding tissues must allow stable geometric construction over the treatment time.

(4) The tumour should not be adjacent to bone—osteoradionecrosis can be a major risk, if sources are placed adjacent to bone.

However, there are also disadvantages to brachytherapy:

(1) The indications are limited and relatively few.

(2) Radiation to personnel.

(3) Anaesthetic is required.

(4) Risk of bleeding and infection.

(5) Hospitalization and isolation required.

(6) Capital costs.

A pre-plan needs to be formulated in order to request the correct amount and activity of iridium wire.

Determine:

(1) The volume to be implanted; Cross-sectional imaging will give additional information to the clinical examination;

(2) The source distribution to be used to cover that; that is the number of sources, their length, and the required separation between them. For instance, for a tongue cancer a common request may be for two or three hairpins, between 12 and 14 mm apart, 45 mm long;

(3) The type of applicator to carry the iridium such as hairpins, wire with tubing, hypodermic needles;

(4) The total implant dose;

(5) The approximate duration of the desired dose-rate for the implant.

The advantages of brachytherapy include:

(1) The short overall treatment time: compared to external beam radiotherapy, the patient is usually in hospital for one week only.

(2) Organ sparing and relative sparing of surrounding normal tissue.

(3) Direct visualization of tumour, and therefore direct placement of radiation into the tumour, unlike the uncertainties associated with external beam radiotherapy. The Gross Tumour Volume (GTV) needs to be accurately demarcated with a small margin for clinical uncertainty to define the Clinical Target Volume (CTV). CTV will equal Planning Target Volume (PTV) as there is no extra margin needed for reproducibility, set up or internal organ movement.

4.2 Specific indications in head and neck cancer

4.2.1 Anterior mobile tongue

Typically, the lesions are well lateralized. Indications are generally restricted to T1 and small T2 cancers. Although some centres have used brachytherapy as a boost following external beam radiotherapy, this is less practised following

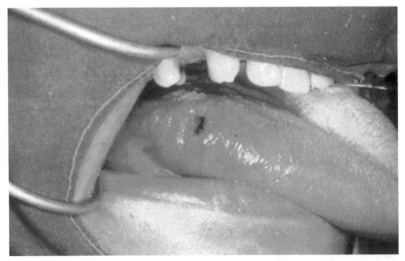

Fig. 4.1 Typical tumour for implantation (see colour Plate 1).

a series of French publications showing brachytherapy alone to be superior to the combination approach. Accurate measurements in the outpatient clinic are vital as they determine the pre-plan (Fig. 4.1).

Generally, the patient is anaethetized using a nasal tube rather than oral. In France, the procedure is often carried out under local anaesthetic in a dental chair. Although a loop technique may be used for thicker tumour, as the indications are usually for small tumours, reliance is primarily on the hairpin technique. It is important to get adequate access with the aid of tooth bites, tongue clips, etc. and then to spend considerable time reviewing the tumour size and marking the GTV accurately with pen or ink.

The guide gutters can be selected for different lengths but normally 60 mm gutters are sufficient. The connecting bridge of the two hollow legs is fixed at 12 mm, which limits the source arrangement using this technique (Fig. 4.2). The guide gutters are placed according to Paris rules to accurately cover the CTV. Intraoperative image intensification is used to ensure the equidistant and parallel geometry. In more posterior tongue cancers, it is important to ensure that the inferior part of the pin does not penetrate the mucosa of the base of tongue. Once a satisfactory position is achieved, a stitch is placed directly underneath the bridge and left open. We use a Rividan needle for this in our institute. Only when the operator is satisfied with the position, does the iridium come into the theatre, and then it is kept shielded in a leaded area, as distance from the sources is important for the radiation protection of theatre staff, taking full advantage of the rapid dose fall off obeying the inverse square law.

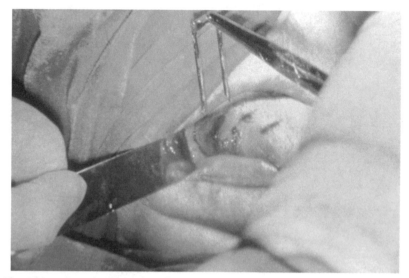

Fig. 4.2 Insertion of guide gutters for iridium hairpins (see colour Plate 2).

Iridium is very flexible (Fig. 4.3). Although this is generally considered an advantage, it is important that when removing the protective foil the iridium is not bent. Some form of guillotine cutter is needed to give an accurate measurement for length. A slight asymmetric cut makes the iridium hairpin easier to guide into the gutters, and the remnants are kept and measured to assist with dosimetry and radiation protection. Grasping the cut hairpin by the bridge with a pair of forceps, and the bridge of the guide gutter with another set of forceps, the hairpins are inserted down the hollow legs of the guide gutters. Once safely in place, the Rividan needle is used to secure the iridium in place in the tongue while the guide gutter is lifted smoothly out of the oral cavity. Further image intensification is required to ensure satisfactory placement. An apparently perfect implant can alter when the thin legs of the iridium no longer have the guide gutters to support them in place. If unsatisfactory the procedure must re-commence. However, repeated trauma to the tongue can cause bleeding and significant oedema (Fig. 4.4).

Much work is being carried out to enhance the dosimetric calculations and optimization with image guidance. CT however is often distorted by the pins, and the oral cavity tumour may be underestimated compared to MR. At present, dosimetrists use orthogonal films to evaluate the implant and calculate the dosimetry using the Paris rules. The basal dose-rate and reference dose-rates are computed and the dosage prescribed to the reference dose-rate. The dose prescribed is generally 65 Gy at a dose-rate of around 10 Gy per day.

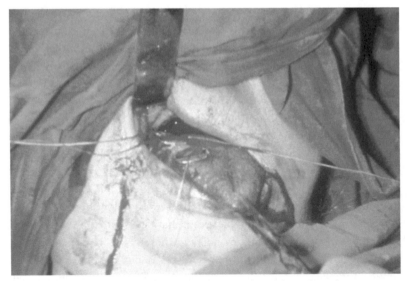

Fig. 4.3 Guide gutters ready for insertion of iridium (see colour Plate 3).

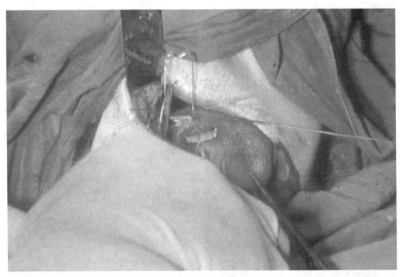

Fig 4.4 Iridium being inserted into guide gutters before their removal (see colour Plate 4).

Care is taken to evaluate areas of potential overdosage. According to the Paris calculations, there are small areas around each leg of the iridium that equate to twice the reference dose-rate. These hyperdosage sleeves may then receive 130 Gy. It is vital to ensure that the implant is as geometric as possible under the Paris rules, to reduce the incidence of soft tissue or bony necrosis.

Mazeron's data described the ideal dose to the reference isodose as 65 Gy. Less than 62 Gy was associated with reduced tumour control, and more than 67 Gy with unacceptable toxicity, in particular, necrosis. For low dose-rate systems, the dose-rate should be between 0.4 and 0.7 Gy per hour. Although the Paris system allows source separation between five and 22 mm, for the oral cavity the ideal source separation is between 10 and 14 mm.

The hairpin technique is similarly used for early stage floor of mouth cancers (Fig 4.5). Proximity to the mandible means that the osteoradionecrosis rate is higher and tumours that abut the mandible are not suitable for primary implantation. The other complication of note is the position of the implant relative to the submandiblular salivary ducts, with the possibility of later sialadinitis. This may cause confusion with malignant lymphadenopathy on follow up.

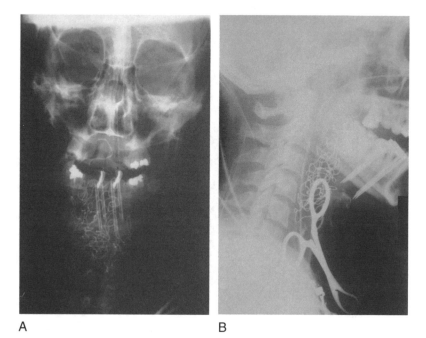

A   B

Fig. 4.5 Anterior (A) and lateral (B) fluroscopic verification of hairpin position prior to insertion of iridium 192.

4.2.2 **Buccal mucosa**

Small tumours of the buccal mucosa lend themselves well to the plastic tube technique. Time is taken in the outpatient's visit to delineate the GTV and calculate the number of tubes, their separation, and the length of tubing required. A single plane implant is the usual, although double plane may be occasionally needed. According to Paris rules, the length needs to take particular account of the requirement for 30% extra length to ensure the adequate active length. Under anaesthesia, the same demarcation of the GTV is achieved with ink on the outer skin surface along with the ideal needle trajectory and length. Care is taken, however, not to introduce the needle directly into the inked area to avoid tattooing. The appropriate length of hollow needle is introduced between the skin and mucosal surface, until the plane is found and then advanced to the proposed exit point. The needles are usually 12 cm or 15 cm long. If a correct plane is found, the needle slides along and small jerky movements with lateral deviations are avoidable. Once all of the needles are in place, the separations and lengths are remeasured. Separations are usually 10 to 15 mm. A solid, stiff plastic cord is then introduced through the hollow needles and the needles themselves are removed. Hollow plastic tubing is cut just longer than the appropriate length, and slid over the stiff plastic. Placing the tubing in hot water for a few minutes makes them more pliable and makes the process much easier. Artery forceps grasp the very tip of the tube and the stiff plastic. In a gentle but firm movement the other end of the cord is pulled so that the plastic cord and the hollow tube combination slips under the skin and sits exactly in place over the marked PTV. The plastic cord is then removed and the damaged clamped end of the cord and tube snipped off. Once the hollow tubing is in place, a marker wire is inserted for the orthogonal films and dosimetry, and temporary bead and lead collar stabilizers are used to ensure that the system stays in place. Later, these are snipped off, so that at an appropriate time the tubes can be manually afterloaded, with the appropriate length of iridium. A few centimetres of blank plastic are symmetrically sealed onto either end of the iridium wire for ease of insertion and removal. The iridium is not brought to the skin surface, to avoid telangiectasia, but sits about 5mm below the surface. The system is then stabilized using small beads and crushable lead clips to ensure that the iridium does not slide out of place. A dose of 65 Gy in 6 to 7 days is usually given. The implant should be checked regularly in situ to ensure no movement of the tubes has occurred, which would critically alter the treated volume. The system is removed after an appropriate length of time by snipping one end medial to the lead cuff and then sliding the tubing and iridium out carefully from the other side. Care must be taken to ensure that the iridium and the tubing come out smoothly together, and iridium is not left embedded in the cheek.

4.2.3 **Base of tongue**

The commonest indication for base of tongue brachytherapy is combined treatment with external beam radiotherapy and where there is nodal involvement with neck dissection also. The tumours should be T1 or T2 and should not have an inferior extent into the vallecula or epiglottis. Other indications may include as single modality in patients who have had recurrence locally post-definitive treatment, or, treatment for small second primaries. A significant number of patients who require recurring treatment are those who have been categorized as patients with squamous cancer neck nodes with unknown primary tumour, and have undergone surgery, and unilateral neck radiotherapy. Eventually, the primary tumour in the base of the tongue becomes apparent, and, because of the previous treatment, extensive surgery is considered the only therapeutic option. Brachytherapy may be considered as a form of organ preservation, if surgery is not possible or unacceptable to the patients. The surgery often amounts to a total glossectomy.

The technique is also a loop technique, but, is technically more difficult, and ideally, should be performed in the presence of a surgeon to ensure adequate placement of the needles over the target volume. The anaesthetized patients are nasally intubated. It is a reassuring practice to palpate the carotid arteries and mark their course on the external skin surface. The outline of the hyoid bone is also often marked on the external skin. The tumour is then palpated internally and the best trajectory from the skin surface to the tumour marked on the skin, ensuring even spacing (usually 10 mm) and a parallel axis. The technique uses two planes and the spacing between the superior and inferior rows of needles is around 12 to 15 mm. The 10 to 15 cm long needles are then inserted until the operator's double gloved fingers palpate their position relative to the tumour. The inferior needles are inserted first followed by the superior row. A stiff plastic cord is then introduced and this exits into the oropharynx. It is then fed down via the superior needle and out again to the skin. It should be ensured that the angle is not too tight, so that the subsequent iridium wire loading is possible. Once the geometry is satisfactory, the needles are withdrawn. The hollow tubing and the marker wire are introduced as described earlier. Particular care is taken when pulling the hollow tube with the stiff plastic cord inside it. It is important to pull firmly but gently into the oropharynx and then feed sufficient tubing in before a second movement from the oropharynx back out onto the skin. Again, the loop should not be too tight. The introduction of marker wire ensures the subsequent passage of the iridium.

The marker wire stays in until the orthogonal films are taken for dosimetric purposes. They ensure the position and conformity to Paris rules. MR scanning may be used to define the length of active iridium required to adequately cover the PTV, and confirm the position of the tubes, relative to the tumour (Fig. 4.6).

As the instrumentation can cause airway threatening oedema, patients are given intravenous steroids and the iridium is inserted only when the airway is absolutely assured, usually 24 hours post-operatively. The beads and lead collar stabilization system is used. Implant removal does not require anaesthetic but needs to be slow and careful so as not to cause discomfort to the patient or produce trauma and bleeding. When external beam radiotherapy is given, a dose of 50 Gy in 25 daily fractions, or its equivalent is given to the tumour, plus margin and nodes, with higher doses to involved nodes. The prescribed dose for the brachytherapy then is around 30 Gy to the reference isodose over 3 days. For relapse or second primary, around 60 Gy is delivered, with dose-rates of 0.3 to 0.45 Gy per hour, if previous therapeutic doses of external beam radiotherapy have been given. Memorial Sloan Kettering Cancer Centre (MSKCC) described their results with this technique, comparing not only local control and survival but also the quality of life as compared to surgery. The local control rates were equivalent, but quality of life was superior with

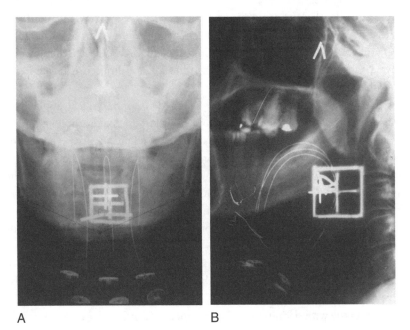

A   B

Fig. 4.6 Orthogonal (A) and (B) radiographs of iridium 192 implantation for base of tongue cancers.

the inclusion of brachytherapy. This plastic tube technique can also be used in tonsil and soft palate cancers.

4.2.4 Lip

Although the same technique can be used for lower lip cancers, a more stable implant uses hypodermic needles. The appropriate length of iridium wire is then fed through the rigid needles directly and secured. Again the length must take into account the Paris rules requiring wires 30% longer than the PTV. The implant should follow the line or the axis of the lip. Pre-formed perforated plastic templates can be used to confer stability and rigidity. The geometric configuration of the source distribution is usually in triangles with 10 mm source separation. This simple technique is cosmetically preferable to surgery or electron beam radiotherapy when a considerable portion of the lower lip is affected either by invasive cancer or by pre cancerous changes. In thicker tumours the implant may be used as a boost following initial electron therapy.

4.2.5 Nasopharynx

High dose-rate brachytherapy is the preferred option in the management of nasopharyngeal cancers. Indications for brachytherapy are limited by the relatively poor depth dose penetration. Patients with nasopharyngeal-confined disease of T1 and T2a may benefit from a boost following external beam radiotherapy plus or minus chemotherapy. In recurrent disease, brachytherapy may have a significant role alongside chemotherapy and limited external beam radiotherapy.

Individual moulds or specifically designed nasopharyngeal applicators may be used, such as the Rotterdam applicator. After adequate preparation of both nostrils using cocaine and adrenaline soaks, a guiding tube is passed via each nostril to exit via the mouth. The Rotterdam applicator then slides over both tubes and is pushed gently into the oral cavity. Forceps are used to ensure the tube does not slip backwards, and forward pressure is applied via the tubes from the nostrils, so that the applicator slips into place in the nasopharynx (Fig. 4.7). When the applicators are correctly positioned, the HDR afterloading catheters are inserted.

The dosimetric calculations are performed and the treatment is then carried out. The calculations may follow the Rotterdam method, described by Levendag, and use orthogonal films. There is an opportunity to CT plan the brachytherapy with definition of the treatment volume and the organs at risk. Boost doses may be fractionated such as 4 Gy in three fractions treating twice daily, or a single boost dose of 7.5 Gy.

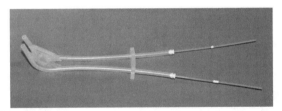

Fig. 4.7 High dose-rate nasopharyngeal brachytherapy.

4.2.6 **Previously treated sites**

Between 5 and 40% of patients with head and neck cancer will develop second primary cancers in the upper aerodigestive tract. Tumours may therefore arise in the previously irradiated sites. Surgical treatment is often offered, but patients with significant co-morbidity may not be considered suitable. Even with modern reconstruction techniques patients may not find the functional results acceptable and wish to explore other options. External beam re-irradiation is sometimes possible, but at a high cost, in terms of side effects. Brachytherapy may be considered, provided the tumour is of small volume and there has been a reasonable disease-free time period. With steep dose fall off, dose to organs at risk, such as the spinal cord, is less of an issue with brachytherapy than with external beam radiotherapy.

Patients with local relapse following maximal treatment may also be considered for further local treatment using brachytherapy. Figures 4.8 and 4.9

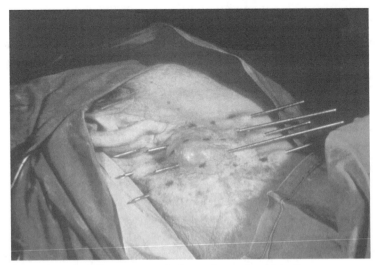

Fig. 4.8 Implantation of recurrent neck disease (see colour Plate 5).

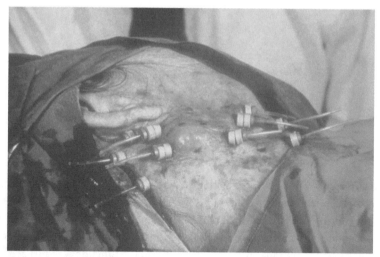

Fig. 4.9 Implantation of recurrent neck disease (see colour Plate 6).

illustrate the use of brachytherapy in a patient with recurrent nodal disease in a previously treated neck. The primary site was controlled and there were no distant metastases. The technique illustrated is similar to that described for buccal mucosa using the principles of the loop technique. Note, in this instance, that it is a two-plane implant; where the basal dose point would be in the centre of triangles drawn under Paris rules. Care must be taken of the underlying carotid artery.

4.3 **Patient care**

Many patients with head and neck cancer have tobacco and alcohol addictions that make radiation protection by isolation difficult. These patients need to have a careful pre-operative assessment and appropriate de-toxification, prior to admission. If there has been trauma during the procedure, then particularly careful attention to the airway is required. The use of high dose intravenous steroids is normally sufficient. If there is any doubt about airway management, the iridium must not be loaded. Patients are nursed in a protected room and they must be relatively self caring. Scrupulous oral hygiene is mandatory, but prophylactic antibiotics are unnecessary. Meticulous skin care on a regular basis with Betadine is required for any puncture sites with the loop techniques. Adequate analgesia initially with a patient controlled system, should include oral non-steroidal anti-inflammatory drugs (NSAIDs), even if only as a gargle. Mild to moderate analgesia is required. In our experience, the patients rarely require intravenous fluids or naso-gastric feeding. The implant must be reviewed carefully at each nurse shift change and there must be a daily review of the patient and verification of the stability of the implant by the dosimetrist.

4.3.1 Complications

Infrequent acute complications include:

(1) Haemorrhage

(2) Infection

(3) Airway distress with oedema

(4) Dental trauma and epistaxis

(5) Sialadenitis of submandibular gland

(6) Poor dosimetric implant requiring removal

Long-term complications include:

(1) Osteoradionecrosis

(2) Soft tissue necrosis—although initially common in tongue implants to a rate of 25%, these normally settle conservatively

(3) Telangiectasia

4.3.2 Results

Anterior tongue and floor of mouth:

Local control rates 90–94%

10% mucosal necrosis

5% osteoradionecrosis for tongue and 20% for floor of mouth.

Lip:

95% local control

Base of Tongue:

MSKCC 5 year survival 75%

Local control T1, T2 100%

Conclusions

Although the role of brachytherapy in head and neck cancer has its limitations, the advantage of preservation of function and cosmesis is extremely important, combined with excellent tumour control. A discussion of the management of the neck is outside of the remit of this chapter. The conformality of the radiotherapy means minimal doses to traditional organs at risk, such as the spinal cord and parotid. There is a definite role for the maintenance of the techniques of brachytherapy in the overall management of head and neck cancer.

Further Reading

Benk V, Mazeron JJ, *et al.* (1990). Comparison of curietherapy versus external irradiation combined with curie therapy in stage 11 squamous cell carcinomas of the mobile tongue: *Radiother Oncol* **18**:229–47.

Gerabulet A, Haie-Meder C, *et al.* (2001). Brachytherapy in the treatment of head and neck cancer.In: Joslin CAF, Flynn A, Hall EJ (ed.). *Principles and practice of brachytherapy.* pp 284–95, London: Edward Arnold.

Levendag P, Lagerwaard FJ, de Pan C, *et al.*(Apr 2002). High dose, high precision treatment options for boost in cancer of the nasopharynx. *Radiother Oncol* **63**(1):67–74.

Mazeron JJ, Simon J, *et al.* (1991). Effect of dose rate on local control and complications in definitive irradiation of T 1-2 squamous cell carcinomas of mobile tongue and floor of mouth with interstitial Iridium 192. *Radiother Oncol* **21**:39–47.

Nag S, Conor ER, *et al.* (2001 Aug 1). The American Brachytherapy Society recommendations for high dose rate brachytherapy for head and neck cancer. *International Journal of Radiation Oncology, Biology and Physics* **50**(5):1190–8.

Chapter 5

Brachytherapy for uterine tumours: cervix and endometrium

Peter Hoskin

5.1 Introduction

Brachytherapy is used in the management of tumours of the uterus in one of two settings:

1. *Post-hysterectomy*

   For most non-bulky stage I and IIA tumours of the cervix, radical surgery will be performed with the removal of parametrial tissue, a vaginal cuff and pelvic lymphadenectomy.

A majority of patients with carcinoma of the endometrium present with early stage I disease, for which the appropriate treatment is total abdominal hysterectomy and bilateral salpingo-oophorectomy. There is debate as to whether pelvic lymphadenectomy be considered a standard treatment in this condition, unlike carcinoma of the cervix, where it is a standard component of the radical hysterectomy procedure.

In such patients, based on specific prognostic features, post-operative radiotherapy may be indicated, within which brachytherapy has an important role.

2. *With the uterus in situ*

   For some more advanced tumours, that is bulky stage IB, II, III, and IV tumours of the cervix, and stage II, III, and IV tumours of the uterus, primary radical radiotherapy will be the treatment of choice and within this brachytherapy is an important component. In this setting there is evidence from population-based studies that the outcome is worse for patients who do not receive brachytherapy.

Occasionally, for early cervical or uterine cancer, brachytherapy alone will be used in particular when the patient is unfit for surgery but the tumour is small and well localized.

5.2 **Early cervical and endometrial carcinoma: post-operative radiotherapy**

Following radical hysterectomy for cancer of the cervix or simple hysterectomy for cancer of the endometrium, certain criteria can be defined, which predict a high risk of residual disease within the pelvis. These are as follows:

Cancer of the cervix:

- ◆ Close or involved surgical margin
- ◆ Lymphovascular invasion
- ◆ More than one pathologically positive lymph node

Cancer of the endometrium :

- ◆ High grade (G3)
- ◆ Stage IC (invasion of outer half of myometrium)
- ◆ Stage II

In these patients, post-operative radiotherapy will be recommended. This will typically take the form of external beam radiotherapy to the pelvis covering the lymphatic drainage of the uterus followed by intravaginal brachytherapy.

5.2.1 **Intravaginal brachytherapy post-hysterectomy**

Both low dose-rate (LDR) and high dose-rate (HDR) systems are available for such treatment, and all use similar applicators. HDR brachytherapy in this setting has the advantage of being a short outpatient treatment distinct from LDR or medium dose-rate (MDR) where an in-patient stay of 20 to 24 hours may be required.

5.2.2 **Vaginal applicator**

The following applicators are available:

(1) Vaginal cylinders of varying circumference, which can be suited to the patient are provided by the commercial brachytherapy manufacturers. It is important to be aware of the individual design of the applicator. Some manufacturers provide specific vaginal stump applicators with the end dwell position close to the surface, whilst some 'standard' vaginal applicators have an end dwell position 5 mm beneath the surface. The importance of this is that, unless stump applicators are used, the surface dose is reduced at the tip of the applicator and hence at the vaginal vault, unless extra weighting is given to the end dwell position. This is shown in Fig. 5.1.

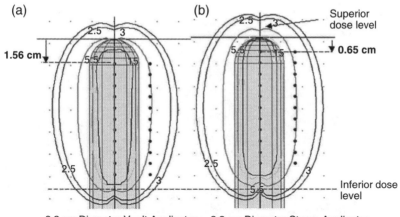

Fig. 5.1 Different dose distributions for vaginal tube (a) and stump applicators (b).

(2) Vaginal ovoids may be used alternatively. The advantage of this approach is to ensure a high vaginal vault dose with good penetration but the disadvantage is that of a more clumsy applicator and reduced proximal dose along the upper vaginal walls.

(3) A further approach is that used at the Institute Gustav Roussy in Paris, where individualized vaginal moulds are made for each patient.

5.2.3 Patient-preparation

LDR systems have the disadvantage of requiring the applicator to be in position for 20 to 24 hours in a typical post-operative setting and therefore patients should be on a constipating regime, for example codeine phosphate 30 mg six hourly, prior to and during the period of treatment.

HDR treatments are performed over a few minutes and therefore, no specific patient preparation is required.

5.2.4 Applicator insertion

In general, applicator insertion can be performed both for LDR and HDR systems without general anaesthetic. For LDR, some system of applicator fixation is required to prevent movement during the lengthy treatment exposure. Commonly a corset-type arrangement is used to fix the applicator in position.

HDR afterloading, is performed as an outpatient procedure within the HDR treatment room and typically, the applicator will be placed in position and held in position by a clamp fixed to the treatment couch for the treatment duration; an example is shown in Fig. 5.2. This duration will typically be 5 to 10 minutes.

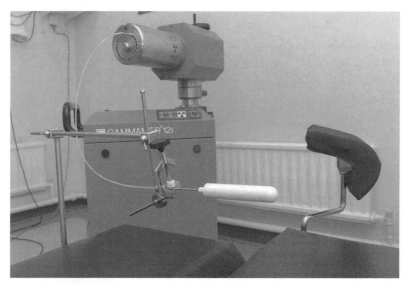

Fig. 5.2 Vaginal applicator clamped in position as for treatment.

5.2.5 **Dosimetry**

The PTV should include the vaginal vault and upper part of the vagina. There is debate as to how much vagina should be included. A standard recommendation would be the upper third of vagina, although some advocate a shorter length.

- If ovoids are used, then inevitably, a dose will only be delivered to the upper 1 to 2 cm of vagina.

- For a vaginal cylinder, a typical treatment length will be 5 cm although where cancer of the cervix is being treated and a vaginal cuff has been removed at radical hysterectomy a shorter length may be appropriate.

Conventionally the dose will be defined at 5 mm from the surface of the applicator, although some centres have adopted a convention which uses the applicator surface dose as the dose prescription point.

Appropriate dwell times are defined using the standard treatment algorithm and software. It should be noted that where a high and deep vaginal vault dose is required then vaginal ovoids will, in general, achieve this more readily than a vaginal stump applicator. Following radical hysterectomy for a cervical carcinoma where the vaginal margin is close then ovoids are more appropriate

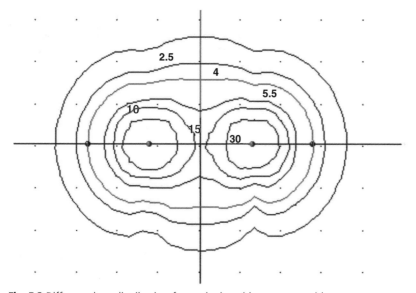

Fig. 5.3 Different dose distribution for vaginal ovoids; compare with stump applicator in Fig. 5.1 (b).

than a vaginal stump applicator, whereas for carcinoma of the endometrium, the main risk is thought to be submucosal lymphatic infiltration, when a vaginal cylinder is best. Examples of the applicators and the differences in dose distribution are shown in Fig. 5.3. It should be noted that uneven dwell times are still required in a vaginal stump applicator to maximize the dose at the end, typical weighting being 2:1.

5.2.6 Treatment implementation

The patient should be catheterized. It is recommended that the treatment be carried out with a full bladder containing at least 100 ml of liquid, which reduces the small bowel volume in the high dose region. This is relatively easy to achieve during the short period of an HDR treatment but more difficult during an LDR treatment when usual practice is for the catheter to be on free drainage.

For the LDR system, the source trains are attached to the applicator. Once this is secure and the patient settled, the treatment will commence. It is important to have some fiducial marker against which to define the position of the applicator and this should be checked on a two-hourly basis during treatment to ensure that the applicator has not moved. Other nursing tasks may also be undertaken at this time.

For high dose treatment, the cylinder or ovoids will be held in position by the couch clamp and the treatment delivered within a few minutes. Significant movement is therefore unlikely as the patient is being observed throughout by means of the CCTV system of the treatment room.

5.2.7 Dose prescription

Following an external beam dose of 40 to 45 Gy in 20 to 25 fractions the following are recommended:

Low dose-rate   15 Gy at 5 mm and 0.5 Gy per hour

Note, this is approximately the radium dose-rate. Most afterloading machines using caesium sources are of medium dose-rate with a dose-rate of 1 to 1.5 Gy per hour and an appropriate total dose correction is therefore needed for this dose-rate. Typically, this will be of the order of 10% delivering a dose of 13.5 Gy at 5 mm. This will mean a treatment time of 22 hours.

High dose-rate   HDR will deliver 11 Gy in 2 fractions.

After an external beam dose of 50 Gy in 25 to 28 fractions, the HDR boost dose will be reduced to 8 Gy in 2 fractions and the LDR boost dose to 10 Gy at radium dose-rate.

The timing of HDR fractions may vary according to machine accessibility but it is assumed that the frequency will not be more than one fraction daily.

5.2.8 Applicator removal

No special technique for removal is required the applicator being readily expelled from the vagina at completion of treatment.

5.2.9 Clinical results

The impact of post-operative radiotherapy in high risk cancer of the cervix has been assumed rather than proven in randomized controlled trials.

In cancer of the endometrium, a clear reduction in local relapse rate has been shown in the Post Operative Radiotherapy in Endometrial Cancer (PORTEC) randomized trial of approximately 10% from 15% without radiotherapy to 2% with radiotherapy in patients with stage IC high grade disease. A similar study is underway in the United Kingdom.

5.3 **Brachytherapy alone post-hysterectomy**

The relative roles of external beam and brachytherapy have not been explored in clinical trials. Similarly, the optimal treatment of the low to intermediate risk patient following hysterectomy for uterine cancer has not been defined. This will include patients with stage IB disease or well differentiated tumours. Some advocate surveillance alone, some external beam treatment and brachytherapy as described above, and others, brachytherapy alone.

Where brachytherapy is given alone, the technique is identical to that above. Prescription differs as follows:

LDR   40 Gy at 5 mm depth at 0.5 Gy per hour dose-rate

An appropriate dose-rate reduction for higher dose-rate sources will be made, for example: 36 Gy at 5 mm delivered at 1 to 1.2 Gy per hour

HDR   22 Gy in 4 fractions at 5 mm depth

Complications from vaginal brachytherapy

In general, this procedure is well tolerated; acute reactions are few, but may include:

- Transient dysuria, often mechanical, from catheterization rather than radiation cystitis.
- Transient proctitis causing bowel frequency and looser consistency

Late complications are:

- Vaginal telangiectasia which may result in occasional blood loss per vagina, particularly after intercourse or clinical examination.
- Vaginal stenosis is seen in over 50% of women with a mean reduction in vaginal length of over 1 cm in the first two years after treatment. Regular use of vaginal dilators can reduce the incidence of stenosis from over 50% to around 10% and should be recommended even in women having regular intercourse.
- An increased incidence in sexual dysfunction is seen and rates of dyspareunia over 40% have been reported, particularly where stenosis is allowed to develop.

5.4 **Cancer of the cervix: no hysterectomy, radical radiotherapy**

The use of intracavitary brachytherapy for the treatment of uterine tumours was one of the first routine applications of this form of treatment. Historically, radium was used and later caesium with a number of different schools evolving, using slightly different applicators and dose-rates, broadly divided into the low dose-rate school of radium use delivering treatment at 0.5 to 0.8 Gy per hour developed in Manchester and the higher dose-rate system developed in Stockholm and Paris with dose-rates of around 1 Gy per hour. When used within the constraints of their applicator design and treatment times they were all effective forms of treatment. Modern gynaecological brachytherapy has evolved from these systems and today there is again a broad division between LDR or MDR systems using cobalt or caesium sources, and HDR afterloading brachytherapy systems using iridium. The general principles of applicator design and insertion technique are, however, common.

Applicator design

Whilst a number of different applicator designs are available, they are all based on the same principle of a central intrauterine tube delivering a dose to the cervix, upper vagina, and parametrial tissues. Whilst a single line source is used by some, most systems include lateral vaginal sources to increase the lateral spread of the dose. Three main types can be considered:

1. Central tube and lateral vaginal sources, often termed ovoids because of their shape. Different eponyms are attached to these applicators according to their specific design and origin; for example the Manchester tube and ovoids, the Fletcher tube and ovoids, where the ovoids are angled posteriorly and have shielding and the Joslin Flynn applicator in which a posterior spatula type shield is included. Examples are shown in Fig. 5 4.

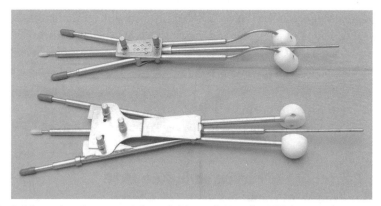

Fig. 5.4 Examples of commercially available applicators for HDR intrauterine treatment using an intrauterine tube and discrete vaginal sources; upper Fletcher Suit applicators, lower Manchester design applicators.

2. Central tube and ring applicator in which the vaginal sources are aligned in a ring beneath the cervix. This is typically used for HDR afterloading, where the sources are small enough to follow the radius of a cervical ring and the presence of varying dwell positions around the ring, distinct from the fixed positions of the ovoids allow for greater individualization of dose distribution where required. Examples of such applicators are shown in Fig. 5.5.

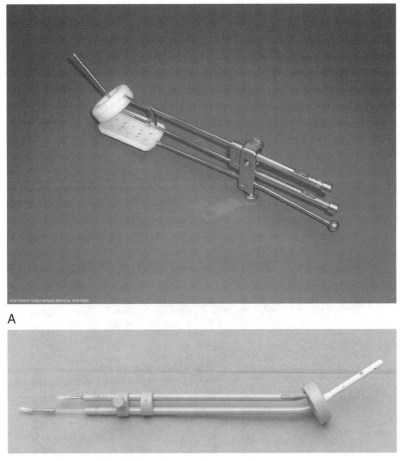

A

B

Fig. 5.5 Tube and ring applicator, with posterior spatula to move rectal wall posteriorly (A) and (B) non-metallic MR compatible applicator without spatula.

3. A single line source is advocated by some for its simplicity and ability to conform to an individualized dose shape by adjusting dwell positions in the single line source. The disadvantage of this is that it delivers dose in three dimensions and so a lateral extension by increasing dwell times would also result in a antero-posterior extension, which will increase dose to bladder and rectum. With such a system, a vaginal tube is typically used to act as a spacer separating the vaginal walls. This system may be chosen preferentially in patients with stage IIA or IIIA disease, where it is necessary to cover proximally down the vagina rather than extend laterally into the parametria. An example is shown in Fig. 5.6.

Patient preparation

Most patients will have had preceding external beam radiotherapy and so attention to control of gastrointestinal upset with appropriate anti-diarrhoea medication, diet, fluid, and electrolyte balance should be carefully assessed.

Many patients with cancer of the cervix run a low grade anaemia and there is good evidence that this impedes the response to treatment. Whilst the role of transfusion is not clear prior to brachytherapy, their haemoglobin should be assessed and maintained above 11.5 g/dl.

Brachytherapy procedure

This is demonstrated schematically in Fig. 5.7.

- In most settings, applicator insertion will be best undertaken under general anaesthetic or spinal anaesthetic. The patient is placed in the lithotomy position and the vulval area cleaned with antiseptic.

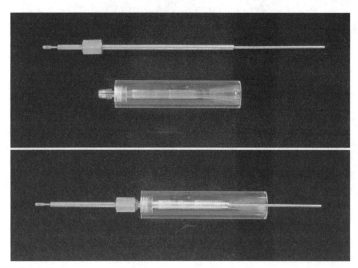

Fig. 5.6 Single line source applicator.

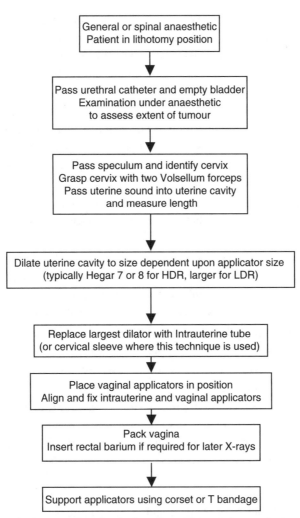

Fig. 5.7 Steps in the procedure for the insertion of intrauterine applicators.

- Examination under anaesthetic should be performed, both per vaginum and per rectum, to assess the clinical extent of tumour and any extension into the vagina or parametrial tissues.

- The cervix should be identified using a speculum grasped with Volsellum forceps. This may be difficult in the case of an advanced tumour where there is extensive necrotic tumour tissue replacing the cervix and it may only be possible to retain adjacent vaginal tissue. This may be adequate, however, to provide fixation of the cervix and some counter-traction for dilatation.

- The cervical canal should be identified using a blunt uterine probe which is passed into the uterine cavity and the length of the cavity can at this point be measured. Where there is extensive necrotic tumour, it may be necessary to gently remove some of this using sponge forceps to identify and access the cervical canal.

- Having identified the cervical canal, this should be dilated with appropriate dilators. The extent of dilatation will depend upon the applicator system to be used; LDR or MDR systems are of larger diameter than HDR systems, the latter often requiring little or no dilatation, while LDR systems requiring the equivalent of a Hegar 7 or 8 dilator to pass.

- Having dilated the cervical canal, if not already inserted, it is best at this point to pass a urinary catheter draining into a urine bag, which will remain for the rest of the treatment.

- Using the dilator, the uterine cavity size can be checked and the intrauterine tube chosen. At this point, it is also necessary to assess the vaginal size for an appropriate vaginal applicator to be chosen. During this time the cervical dilator should be retained in the canal to prevent it from closing down.

- The dilator should then be replaced by the intra-uterine tube followed by the vaginal source. Different applicator systems have different forms of clamp or fixation of the two or three applicators together so that a rigid geometric relationship is achieved for each insertion.

- Gauze packing is then inserted into the vagina posteriorly to displace the rectum and prevent movement of the applicators. Alternatively, an applicator system using a posterior spatula, for example the Joslin Flynn applicator can be used. Care should be taken so as not to pack behind the vaginal source and displace them in a caudal direction. If plain orthogonal films are to be taken, 5 to 10 ml of barium should be placed in the rectum; However, this could cause considerable artefact on CT and so is better omitted if CT is to be used.

- The insertion is then complete. For LDR or MDR systems, an anchoring corset is often used, as these applicators will have to stay in place for many hours. For HDR systems, less rigid fixation is necessary and a T bandage or elasticated bandage is sufficient.

- Verification imaging is a vital component of the procedure. As a minimum, this should be AP and lateral orthogonal X-ray films with a magnification marker; ideally, CT imaging will be undertaken and wherever MR compatible applicators are available, the soft tissue definition is undoubtedly superior with MRI.

- Where HDR is used, imaging is essential before each fraction.

Cervical sleeve technique

Because of the need for fractionated treatment with HDR and to avoid repeated procedures under anaesthetic where HDR is to be used, a cervical sleeve technique may be employed. This entails, prior to insertion of the intra-uterine tube and following dilatation of the cervical canal, insertion of a plastic sleeve retained by sutures onto the cervix or where there is extensive tumour sutured to the adjacent vaginal walls. This then acts as a conduit for the cervical tube and allows further insertions as an outpatient procedure. An example of such a sleeve is shown in Fig. 5.8.

Other special considerations

◆ *Haematocolpos* may be encountered secondary to cervical canal stenosis and obstruction. This will be drained at the time of dilatation and provided there is no evidence of infection the insertion can proceed.

◆ *Pyocolpos* is a more common complication, often unsuspected until pus is released at the time of cervical dilatation. The pus should be drained off during the procedure and treatment with broad spectrum antibiotics be commenced, after taking swabs for culture.

◆ If an LDR/MDR insertion is planned this should be abandoned and attempted again following a course of antibiotics; some also recommend a drain to be placed within the uterine cavity to ensure continued drainage of pus.

◆ If an HDR procedure is planned, this can proceed. If a cervical sleeve is to be inserted, provided it is of a design having drain holes in the end, then this can also proceed, acting as a drain for any residual pus or fluid.

Fig. 5.8 Cervical sleeve.

- *Fistulae* secondary to tumour invasion may be encountered at examination under anaesthetic (EUA); they are not in themselves a specific contraindication to intrauterine brachytherapy. If urine leakage is suspected, then instillation of methylene blue into the bladder may aid identification of leakage per vagina.

- *Perforation* may occur even with the most skilled operator. Typically, this is through the posterior wall of the cervical canal into the pouch of Douglas, or less often, through the fundus. Series using CT scanning have suggested that the true rate of perforation, often unsuspected clinically, may be as high as 5%. If perforation is suspected, unless the canal can be defined with confidence and the perforation bypassed, the procedure should be abandoned and the patient observed for 24 hours for signs of fever or peritonism.

Implant dosimetry

Gynaecological brachytherapy is undertaken following as closely as possible the recommendations in the ICRU Report No. 38. This builds on the long tradition of gynaecological brachytherapy, which defines doses at named specific points, and also includes the concept of defining a 60 Gy isodose envelope.

Conventional dosimetry will be based on orthogonal antero-posterior and lateral X-ray films with the applicators in situ. The rectum is defined using barium and the bladder from the balloon of the bladder catheter containing 7 ml of radio-opaque liquid.

The standard dose points are then defined as follows:

POINT A: 2 cm lateral to the midline, and 2 cm above the surface of the ovoid in the lateral vaginal fornix.

POINT B: 3 cm lateral to point A.

ICRU bladder reference point, which is at the inferior part of the bladder catheter balloon.

ICRU rectal reference point, which is on the anterior rectal wall at a point perpendicular from the cervical os or lowest vaginal source.

These are shown in Fig. 5.9.

These reference point doses are the absolute minimum which should be defined in cervical cancer. It is preferable to produce the full isodose distribution and is ideal to undertake CT imaging and define dose volume histograms.

A further development is the use of MR compatible applicators with which an accurate PTV can be defined from the MR definition of the tumour.

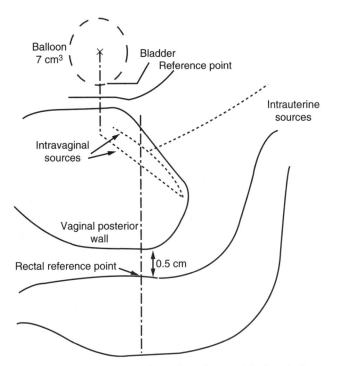

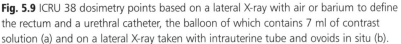

Fig. 5.9 ICRU 38 dosimetry points based on a lateral X-ray with air or barium to define the rectum and a urethral catheter, the balloon of which contains 7 ml of contrast solution (a) and on a lateral X-ray taken with intrauterine tube and ovoids in situ (b).

Dose prescription

Conventionally, the dose prescription is defined at point A. It is, however, important to realize that dose definition at a point is not the full description of the dose distribution and the same point dose can be achieved with varying distributions around that point. Furthermore, in evaluating the published literature, it is important to realize that different authors and brachytherapy groups have re-defined point A in slightly different ways, as shown in Fig. 5.10.

Another general principle by which dose schedules are derived is that, where the tumour is confined centrally, bearing in mind that point A will encompass 2 cm radius around the cervical canal, relatively more useful dose can be given by brachytherapy, but when there is tumour extending outside this 2 cm radius, external beam treatment has a more important role. For these reasons there is a wide range of individual prescriptions which can be found within cancer centres across the world.

Whilst in principle, the dose prescription is simple, in practice it has caused confusion and in some instances disaster because of the importance of considering dose-rate.

The conventional dose prescription is to deliver 75 to 85 Gy to point A. This dose may be composed of external beam treatment and brachytherapy. Dose-rate of the external beam component will be that of a linear accelerator at approximately 1 Gy per hour and typical schedules will deliver 45 to 50 Gy in this way. The remainder will then be delivered by brachytherapy. The complexity and confusion then relates to the fact that the excess dose, for example if 50 Gy is given by external beam the additional 25 to 35 Gy, is defined at radium dose-rate by 0.7 Gy per hour. This is no longer used in clinical practice and the medium dose-rate machines using a caesium source will deliver a dose at around 120 cGy (1.2 Gy) per hour. If a dose of 25–35 Gy is given without

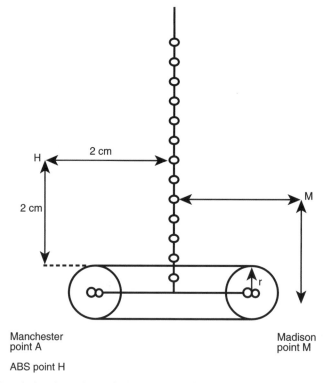

Fig. 5.10 Variations in 'Point A' dosimetry points for cervical cancer brachytherapy.

correction for dose-rate, an excess biologically effective dose is delivered with potential disaster for normal tissue.

An HDR afterloading iridium source delivers dose at approximately 1 Gy per minute with some variation from source decay during a typical period of use of around three months. When HDR systems are considered, fractionation is required.

Meta-analysis comparing LDR with HDR suggests that the ultimate clinical result is very similar. The following rules of thumb should be used and allow safe brachytherapy delivery:

◆ For MDR of around 1 to 1.2 Gy per hour, a 10 to 15% dose reduction from the predicted radium dose should be used.

◆ For HDR brachytherapy, a 55% dose reduction should be used and fractionation employed such that the dose per fraction does not exceed 7 Gy.

Common schedules in use in the United Kingdom therefore would deliver 50 Gy external beam and follow this with:

◆ 20 to 22 Gy MDR or

◆ 14 Gy in 2 fractions HDR.

In other parts of the world, higher doses are often given and a summary of the recommendations of the American Brachytherapy Society are shown in Tables 5.1–5.3. Note external beam therapy in these schedules often uses rectal shielding to enable higher brachytherapy doses.

A further consideration needed to keep the total dose within normal tissue tolerance is the critical organs at risk being the rectum and the bladder. Total equivalent doses to the rectum of >65 Gy and to the bladder of >70 Gy will result in an increased incidence of late damage. A useful practical rule of thumb is to keep the rectal dose to no more than two-thirds of the point A dose. Where this is exceeded, a clinical decision must be taken as to whether the brachytherapy dose should be curtailed to remain within tolerance.

Table 5.1 ABS recommendations for advanced cervical cancer

External beam dose	*LDR Point A dose	HDR Point A dose
45–50 Gy	40 Gy	4 × 7 Gy
		5 × 6–6.5 Gy
		6 × 5.8–6 Gy

*LDR dose-rate of 0.5–0.65 Gy per hour

From: Nag S et al. *International Journal of Radiation Oncology Biology Physics* 2000; 48: 201–211.

International Journal of Radiation Oncology Biology Physics 2002; 52: 33–48.

Table 5.2 ABS recommendations for post-operative radiation therapy for cervical cancer

External beam dose	*LDR surface vaginal dose	HDR surface vaginal dose
45–50.4 Gy		
Microscopic positive vaginal margin	20 Gy	No recommendations
Gross residual tumour	30–35 Gy	

*LDR dose-rate of 0.5–0.65 Gy per hour

From: Nag S et al. *International Journal of Radiation Oncology Biology Physics* 2002; 52: 33–48.

Table 5.3 ABS recommendations for post–operative HDR brachytherapy in endometrial cancer; any of the doses below are considered acceptable

HDR alone	After 45 Gy external beam (1.8 Gy fractions)
3×7 Gy at 0.5 cm	2×5.5 Gy at 0.5 cm
3×10.5 Gy surface dose	2×8 Gy surface dose
4×5.5 Gy at 0.5 cm	3×4 Gy at 0.5 cm
4×8.8 Gy surface dose	3×6 Gy surface dose
5×4.7 Gy at 0.5 cm	
5×7.5 Gy surface dose	

From: Nag S et al. *International Journal of Radiation Oncology Biology Physics* 2000; 48: 779–790.

Treatment delivery

Low to medium dose-rate These treatments will require hospitalization and isolation in a purpose built afterloading brachytherapy room. During treatment, it is important that applicator position is verified and the treatment may be interrupted for this and other nursing tasks. A fiducial mark preferably on the patient in the form of an indelible skin mark, adjacent to a similar mark on the applicator, should be used to check movement of the applicators two hourly through treatment.

High dose-rate HDR treatment will be delivered in the HDR treatment room with the applicators connected using the appropriate source tubes. Typical treatment time is of the order of 8 to 10 minutes, depending upon the source activity.

Applicator removal

Removal is undertaken in the treatment room, whether the LDR, MDR after-loading room, or the HDR treatment room. Following removal of any bandaging support and the gauze vaginal packing the applicators will generally fall out without difficulty. The bladder catheter should also be removed at this time.

Clinical results

Stage IB: Radical radiotherapy utilizing brachytherapy is highly effective with equivalent cure rates to those achieved with radical surgery and overall five year survival figures of 85%.

More advanced stage disease will have a worse prognosis with 60 to 70% of patients with stage II carcinoma of the cervix, surviving five years or more, and only 30 to 40% of those with stage IIIB achieving this.

It has however been shown that patients receiving brachytherapy have a much better prognosis than those who do not.

Treatment complications

Complications directly attributable to the brachytherapy relate to the very high dose delivered around the cervix, upper vagina, bladder base and anterior rectal wall. These may be considered as follows:

◆ Vaginal side effects are predominantly related to stenosis and shortening of the vagina which can be prevented to some degree by the use of vaginal dilators.

◆ Rectal complications will be compounded by the external beam dose delivered also and include rectal frequency and bleeding due to telangiectasia. Severe rectal problems should be seen in no more than 5% of patients but less troublesome grade I and II side effects are seen in over 30%.

◆ Bladder side effects will include frequency and haematuria from bladder telangiectasia. Occasionally, urethral structuring may also develop requiring dilatation. These will typically be seen in less than 5% of patients.

◆ Rectal complications develop sooner than bladder complications the mean time to onset for rectal complications being two to three years, with bladder problems developing on average a year or two later.

5.5 **Intact uterus, carcinoma of the endometrium**

Endometrial cancer is undoubtedly best managed by primary hysterectomy when localized to the uterus. There will be instances however where patients are unfit for hysterectomy particularly, as this is a tumour of the obese and related to diabetes and hypertension in its population. In around 20% of patients there will be spread outside the uterus and these patients, will also be treated with radical radiotherapy in preference to surgery. The indications for brachytherapy in endometrial carcinoma are therefore as follows:

◆ Localized tumour in the medically unfit.

◆ Advanced tumour extending beyond the uterus, in combination with external beam radiotherapy.

5.5.1 **Applicators**

In some cases the same applicators are used as those for cervical cancer, either a central intrauterine tube alone or with two vaginal sources. This has, however, been considered less than satisfactory in the past before the flexibility of afterloading machines, which enables varying and individualized dose distributions to be delivered. Historically, therefore, a number of other applicator systems have been developed taking account of the fact that in endometrial cancer, the uterine cavity is often enlarged and dilated and a central intra-uterine tube therefore may not be able to deliver dose to the full thickness of the myometrium without careful planning, despite which a high surface dose may be necessary. Examples include the following:

◆ Heyman's capsules were developed in Stockholm for the Stockholm system using caesium pellets. These comprise individual capsules which are packed into the uterine cavity as shown in Fig. 5.11. This enables a high activity to be concentrated within the endometrial cavity extending through the myometrium.

Heyman's capsules are still used by some who contend that they give the best distribution within the intrauterine cavity. Commercial systems are available for the common HDR afterloading machines.

◆ An alternative is to use two or three intrauterine tubes to cover the full extent of the dilated cavity and such applicators are available commercially. A spring loop containing caesium sources has also been used introduced into the uterine cavity to again ensure the entire volume was treated. The type of applicator will to some extent be determined by the individual situation which is being treated.

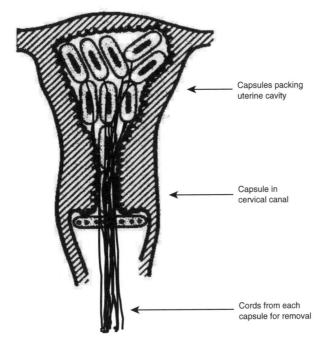

Fig. 5.11 Heymann's capsules

- For early disease, either a single or multiple intra-uterine tube applicator with appropriate adjusted dwell times or Heyman's capsule, will be most appropriate.
- For advanced disease extending outside the uterus, a more standard applicator such as that used for cervical cancer enabling dose to be increased laterally with the vaginal source, either a tube and ovoids or tube and ring applicator system may be more appropriate.

5.5.2 **Patient preparation**

This will be the same as for intrauterine insertions for cervical cancer. Specific considerations for those with endometrial cancer are as follows:

- Co-morbid medical condition is much more common and attention towards blood pressure control and diabetes management is important.
- Perhaps as a consequence of the above, thrombosis as a post-operative complication is a recognized feature, particularly where LDR or MDR systems are used requiring prolonged immobilization. Consideration should therefore be given to thrombosis prophylaxis with elasticated stockings and subcutaneous heparin.

5.5.3 **Applicator insertion**

In general, this will follow the same steps as that for cervical cancer with the following modifications:

◆ The uterus is often enlarged but the cervix will be normal and therefore, identification and dilation of the cervical canal is usually easier.

◆ Where a single source is to be used, the intra-uterine tube is inserted as usual and a vaginal applicator is placed over it to aid its fixation and separate the vaginal walls. Packing is not required with this type of applicator but suturing of the tube at the introitus is recommended to prevent displacement.

◆ Heyman's capsules are inserted one by one into the cavity until it is packed, the number varying on an average between nine and twelve capsules. It is important that the final capsule is placed in the cervical canal to keep it dilated and to enable removal of the capsules at completion of treatment.

5.5.4 **Dose prescription**

The same constraints with regard to dose-rate apply to the treatment of uterine endometrial carcinoma. Whilst the ICRU 38 guidelines will still be used and the dosimetry points identified, the dose distribution may be individualized. Uterine endometrial cancer requires relative loading of the dose into the more distal parts of the intra-uterine tube to ensure coverage of the fundus and uterine wall thickness. Weighting of sources distally, therefore, is often required. The preferred method is to undertake CT scanning to define the uterine width and hence CTV which will be to the outside of the uterine wall and this is also the PTV for the brachytherapy volume prescription. With afterloading, an individualized dose distribution can then be defined as shown in Fig. 5.12.

The dose prescription will depend upon the combination with external beam treatment.

For stage I disease, in medically unfit patients, brachytherapy alone may be considered appropriate, in which case the dose prescription should be:

◆ LDR/MDR: 75 to 80 Gy to point A or the defined tumour volume

◆ HDR: 36 to 42 Gy in 6 fractions to point A or the defined tumour volume (PTV)

As in all circumstances, the final dose delivered should be within normal tissue constraints and in particular, the rectal dose should not exceed two-thirds of the tumour prescription dose.

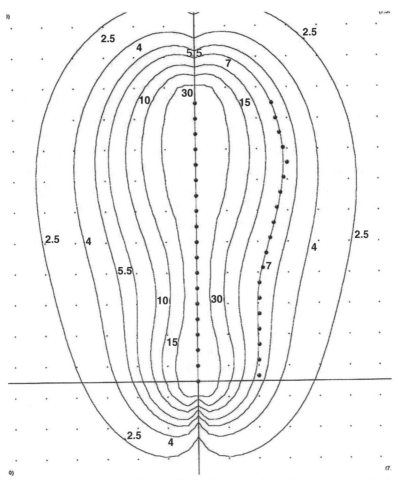

Fig. 5.12 CT defined optimized dose distribution for endometrial cancer using single intrauterine tube; prescription isodose is 7 Gy.

Where external beam radiotherapy is delivered the prescription will be the same as for cervical cancer; in other words 50 Gy delivered by external beam followed by 25 to 35 Gy to point A at 0.7 Gy per hour and therefore with appropriate dose-rate correction with MDR/LDR 22 to 25 Gy at 1 to 1.2 Gy per hour or HDR fractionated treatment as described under cervical cancer. The most usual prescription in the United Kingdom is 14 Gy in 2 fractions and in the United States 24 Gy in 4 fractions.

5.5.5 **Clinical results**

Results of treatment of early stage disease are broadly comparable with surgery, although large data sets have not been published. This reflects the fact that this treatment is only chosen for the medically unfit.

For more advanced disease, that is stage III or IVA, then the outlook is for a 30 to 40% five year survival with stage III disease and 10% or less with stage IV.

5.5.6 **Tumour complications**

These are essentially those described under cancer of the cervix.

As mentioned earlier, these patients are medically unfit, often obese with co-morbid hypertension, and diabetes, and therefore the risk of medical complications, and in particular deep venous thrombosis is high.

Further reading

Gerbaulet A, Potter R, Haie-Meder C (2002). Cervix cancer. In: Gerbaulet A, Potter R, Mazeron JJ, Meertens H, Evan Limbergen (eds.). *The GEC ESTRO handbook of brachytherapy*, pp 301–64. Brussels: ESTRO.

Hunter RD, Davidson SE (2001). Low dose-rate brachytherapy for treating cervix cancer. In: Joslin CA, Flynn A, Hall EJ (ed.). *Principles and practice of brachytherapy using afterloading systems*, pp 343–53. London: Arnold.

ICRU Report 38 (1985) Dose and volume specification for reporting intracavitary therapy in gynecology. International Commission of Radiation Units and Measurements, Bethesda Maryland, USA.

Jones BM, Pryce PL, Blake PR, Dale RG (1999). High dose-rate brachytherapy practice for the treatment of gynaecological cancers in the UK. *British Journal of Radiology* 72:371–77.

Joslin CF (2001). High dose-rate brachytherapy for treating cervix cancer. In: Joslin CA, Flynn A, Hall EJ (ed.). *Principles and practice of brachytherapy using afterloading systems*, pp 354–72. London: Arnold.

Ladner HA, Pfleidereer A, Ladner S, Karck U (2001). Brachytherapy for treating endometrial cancer. In: Joslin CA, Flynn A, Hall EJ (ed.). *Principles and practice of brachytherapy using afterloading systems*, pp 333–42. London: Arnold.

Nag S , Erickson B, Thomadsen B, Orton C, Demanes JD, Petereit D for the American Brachytherapy Society (2000). The American Brachytherapy Society recommendations for high-dose-rate brachytherapy for carcinoma of the cervix. *International Journal of Radiation Oncology, Biology and Physics* 48:201–11.

Nag S, Erickson B, Parikh S, Gupta N, Varia M, Glasgow G for the American Brachytherapy Society (2000). The American Brachytherapy Society recommendations for high-dose-rate brachytherapy for carcinoma of the endometrium. *International Journal of Radiation Oncology, Biology and Physics* 48:779–790.

Nag S, Chao C, Erickson B, Fowler J, Gupta N, Martinez A, Thomadsen B for the American Brachytherapy Society (2002). The American Brachytherapy Society recommendations for high-dose-rate brachytherapy for carcinoma of the cervix. *International Journal of Radiation Oncology, Biology and Physics* 52:33–48.

Potter R, Gerbaulet A, Haie Meder C(2002). Endometrial cancer. In: Gerbaulet A, Potter R, Mazeron JJ, Meertens H, Evan Limbergen (eds.). *The GEC ESTRO handbook of brachytherapy*, pp 365–402. Brussels: ESTRO.

Thomas R and Blake P (1996). Endometrial Carcinoma: Adjuvant Locoregional Therapy. *Clinical Oncology* **8**:140–45.

Chapter 6

Prostate cancer: permanent seed brachytherapy and high dose-rate afterloading brachytherapy

Dan Ash, Peter Hoskin

6.1 Introduction

Permanent seed implantation for prostate cancer is essentially for patients with localized disease who have a high probability of cure. There are no specific symptoms caused by localized prostate cancer and many patients present with symptoms of benign prostatic hypertrophy, which is investigated by performing a PSA. There is also an increasing group of asymptomatic patients who request PSA for screening. A raised PSA is an indication for transrectal ultrasound and biopsy to confirm the diagnosis.

6.2 Staging (TNM System 1992 Revision)

T1A: less than 5% TURP chippings positive for prostate cancer.

T1B: more than 5% TURP chippings positive for prostate cancer.

T1C: impalpable prostate cancer found by biopsy.

T2A: palpable nodule involving less than one half of the prostate lobe involved.

T2B: palpable nodule involving more than one half of the prostate lobe involved.

T2C: bilateral nodules or one nodule involving both lobes.

T3:   tumour invading outside the prostate capsule into bladder neck or seminal vesicle but not fixed.

T4:   tumour fixed to invading adjacent structures.

Permanent seed implantation is only suitable for patients in stages T1 and T2.

6.3 **Investigation**

The following investigations are mandatory:

1. clinical history;
2. digital rectal examination to evaluate clinical stage;
3. PSA;
4. transrectal ultrasound to assess local extent and prostate volume;
5. prostate biopsy to confirm adenocarcinoma with Gleason score;

Patients with a PSA of less than 10 and a Gleason score of 6 or less are very unlikely to have the disease outside the prostate. For patients with a PSA of greater than 10 and/or a Gleason score of 7 or more an MRI scan is recommended to assess the local extent of the disease and to evaluate pelvic nodes.

Bone scan should only be considered in those patients with a PSA of greater than 10.

6.4 **Indications and contra-indications**

The aim is to select patients who have a high probability of disease confined within the prostate capsule where cure is likely and to select those who will have a good functional outcome.

The main prognostic factors for cure are PSA, Gleason score and stage. Patients with a good prognosis, i.e. PSA less than 10, Gleason score less than 6, stage T2A or less are suitable for brachytherapy alone.

Patients with an intermediate prognosis, i.e. PSA greater than 10 or Gleason score greater than 7, can also be treated by brachytherapy alone, but as the prognostic factors worsen, the risk of disease outside the prostate capsule increases and there may be an indication for combining brachytherapy with external beam radiation for some of these patients. The factors which increase risk include the number of positive biopsies, the proportion of biopsy tissue invaded, and the presence of perineural invasion.

High-risk patients have both a PSA of greater than 10 and a Gleason score of 7 or more. These patients have a fairly high risk both of extra-capsular spread and metastatic disease. They are not suitable for permanent seed brachytherapy alone but some can be treated by a combination of external beam radiation with brachytherapy used as a boost.

In addition to defining a population of patients likely to be cured it is also necessary to define those who are likely to have a good functional outcome without too great a risk of side effects. This is related to the state of the urine outflow before treatment. Patients with large prostates (greater than 40 cm^3) and significant flow symptoms as determined by the International Prostate

Symptom Score (IPSS) are at higher than average risk of going into acute retention after brachytherapy and having prolonged side effects.

If the prostate volume is greater than 50 ccs, it is difficult to achieve a satisfactory implant because some of the prostate is often shielded by the pubic arch. For patients who are otherwise suitable three months of neo-adjuvant hormone therapy will usually produce a 30% reduction in prostate volume and make them suitable for implantation.

Patients who have had a recent TURP should not be implanted because it is difficult to achieve a satisfactory dose distribution and there is a high risk of incontinence after treatment. If TURP has been performed several years ago and the cavity has regrown, then brachytherapy can be considered, though the risk of incontinence is still slightly higher than in case of those who have not had a TURP.

6.5 Isotopes for permanent seed implantation

6.5.1 Iodine 125

The most frequently used isotope for permanent seed implantation is iodine 125. It has a mean energy of 25 KeV with a half life of 59.4 days (see Chapter 2).

6.5.2 Palladium 103

Palladium has a half life of 17 days with a mean energy of 27 KeV. It was initially thought that palladium might be preferable for the more aggressive tumours with higher PSA and Gleason score because of its more rapid dose-rate. Subsequent clinical experience suggests that the outcome for palladium and iodine is the same (see Chapter 2).

6.6 Implant procedure

6.6.1 Volume definition

As with all other forms of brachytherapy, it is important to know the volume of the prostate to be implanted and the position and number of seeds to be used before starting the treatment. Most centres use low activity seeds (0.4 mCi per seed). This usually produces a seed density of approximately 2.5 seeds per cubic centimetre. The target volume is the whole of the prostate as defined by the capsule plus a margin of 2 to 3 mm. The seminal vesicles are not usually included within the target volume.

Before implantation the patient has a transrectal ultrasound volume study. This is performed in the lithotomy position with a transrectal ultrasound probe inserted into the rectum. The probe is placed within a stepping unit which can incrementally advance and retract the probe within the rectum and is fixed to the floor. Attached to the transrectal ultrasound probe is a template. The coordinates of the template are automatically transposed over the ultrasound images of the prostate.

The prostate is positioned so that it lies centrally within the template grid with the lower border on the first row and the urethra centred in the middle row. Care should be taken not to have the prostate angled or rotated around the axis of the probe.

Once the prostate is accurately positioned relative to the template serial ultrasound sections are taken from the base to the apex at 5 mm intervals as shown in Fig. 6.1. On each section the prostate capsule is outlined and the data is inserted into a planning computer to calculate the exact number and position of seeds required for the implant as shown in Fig. 6.2.

It is possible to perform the planning and implant under the same anaesthetic provided the volume of the prostate is known, and to create the dose plan interactively as the implant proceeds. This is done by recording the position of each seed within the implanted volume and building up the isodoses as the implant proceeds. This ensures that full coverage of the prostate capsule is achieved by the treatment isodose at the end of the implant, using a modified peripheral loading pattern.

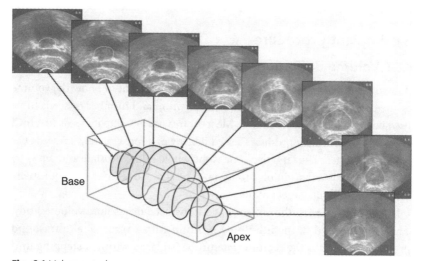

Fig. 6.1 Volume study.

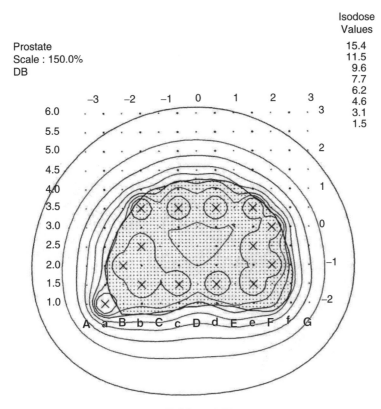

Level : 6 Z-Offset : 2.50

Fig. 6.2 Isodose plan.

6.6.2 **Implantation of sources**

Accuracy is ensured by using transrectal ultrasound guidance as shown in Fig. 6.3 (overleaf). Implantation is performed by passing a 20 cm long 18 gauge needle through the template and perineal skin into the prostate. The template provides the X and Y coordinates and the depth or Z coordinate is given by the treatment plan. Not all needles are inserted to the same depth. The depth of the needle is referred to the base plane where the prostate meets the bladder base. Needles destined for the prostate apex may be 3 to 4 cm caudal to the base plane.

The position of the needle at depth is determined by setting the transrectal ultrasound plane at the required distance from the base plane and then guiding the needle into the X and Y coordinates until it reaches the plane at

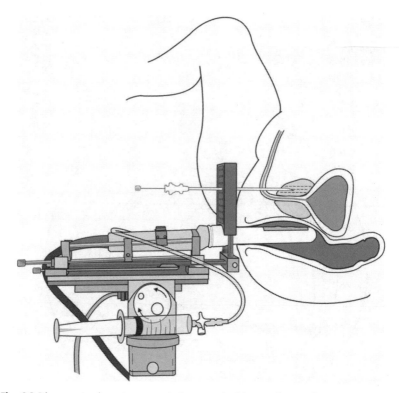

Fig. 6.3 Diagram to show transrectal ultrasound with template and needle being inserted into the prostate.

the correct depth. The seeds are then inserted into the needle either through a Mick applicator or by using pre-loaded strands of 5 mm spaced linked seeds.

A typical implant of a 40–50 ml prostate will use 25 to 30 needles and place 80–100 seeds within the treated volume.

6.7 **Prescription dose**

For Iodine 125 the aim of the implant is to deliver a minimum peripheral dose of 145 Gy to the whole prostate plus a margin of 2 to 3 mm. Approximately 40 to 60% of the implanted volume will receive 50% more than the prescribed dose (220 Gy). About 20% will receive 200% of the dose (290 Gy).

For patients who are receiving iodine seed brachytherapy as a boost after 50 Gy external beam radiation the dose is reduced to 110 Gy.

Because of the higher dose-rate for palladium 103 implants the dose is reduced to 125 Gy if patients are treated by brachytherapy alone and to 100 Gy if brachytherapy is delivered as a boost after external beam radiation.

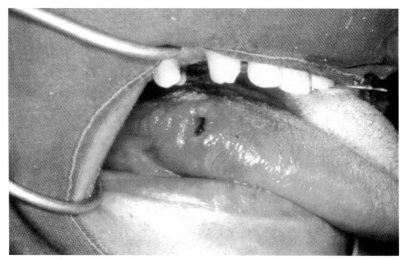

Plate 1 Typical tumour for implantation (Fig. 4.1 in text).

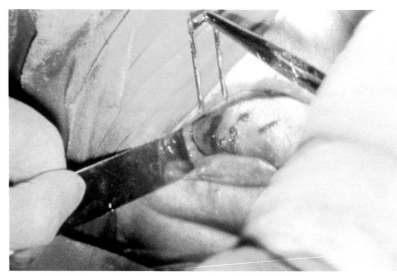

Plate 2 Insertion of guide gutters for iridium hairpins (Fig. 4.2 in text).

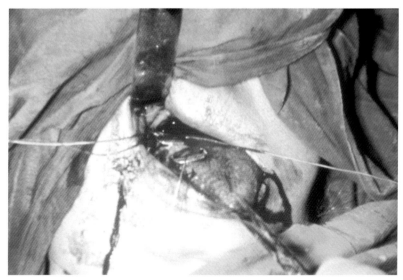

Plate 3 Guide gutters ready for insertion of iridium (Fig. 4.3 in text).

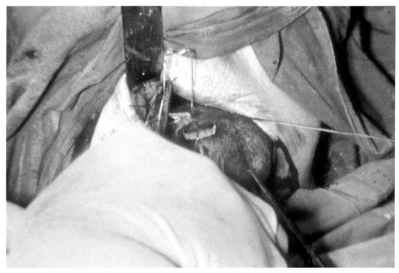

Plate 4 Iridium being inserted into guide gutters before their removal (Fig. 4.4 in text).

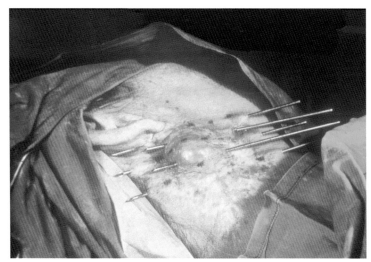

Plate 5 Implantation of recurrent neck disease (Fig. 4.8 in text).

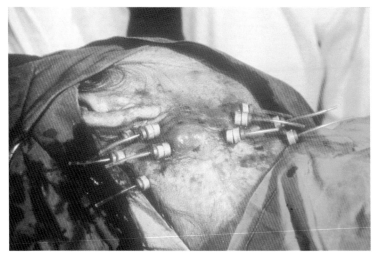

Plate 6 Implantation of recurrent neck disease (Fig. 4.9 in text).

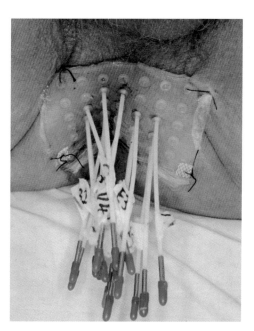

Plate 7 HDR prostate implant using flexible implant system in situ (Fig. 6.5 in text).

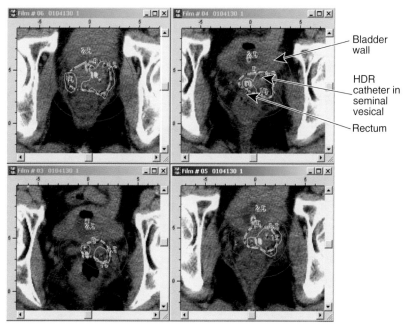

Bladder wall

HDR catheter in seminal vesical

Rectum

Plate 8 CT planning scan of HDR implant showing implant extended to include seminal vesicles (Fig. 6.6 in text).

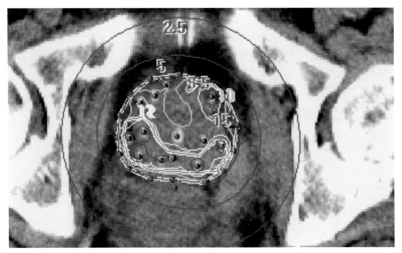

Plate 9 Typical dose distribution from HDR implant (Fig. 6.7 in text).

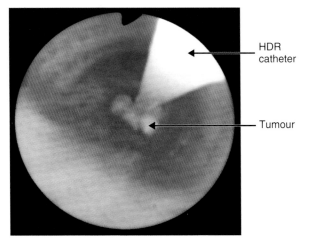

HDR
catheter

Tumour

Plate 10 View of endobronchial tumour and HDR catheter passed beyond it through bronchoscope (Fig. 7.1 in text).

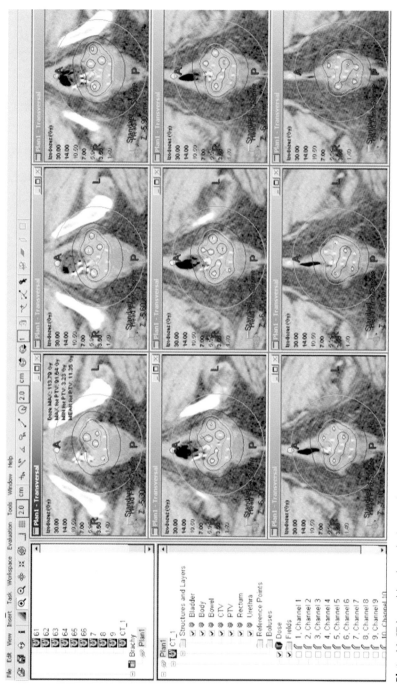

Plate 11 CT based implant dosimetry (Fig. 8.3(a) in text).

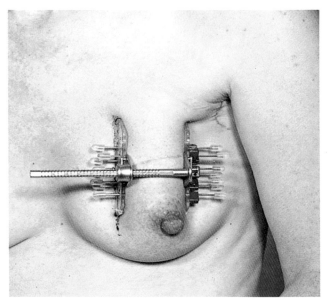

Plate 12 Breast implant in situ (Fig. 9.2 in text).

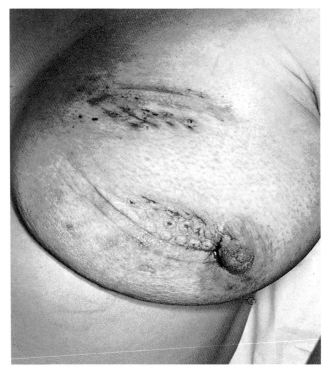

Plate 13 Breast appearance immediately following implant removal with oedema and indentation from bridge sites against skin (Fig. 9.5 in text).

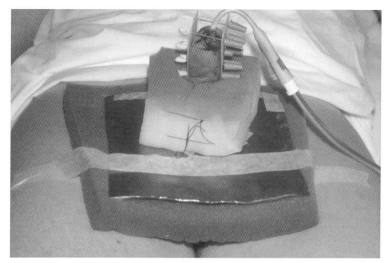

Plate 14 Penile implant scaffold effect (Fig. 10.2 in text).

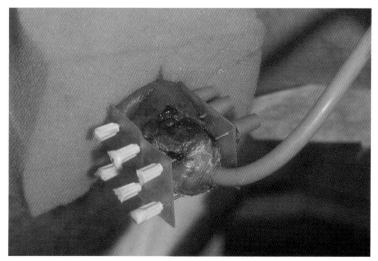

Plate 15 Penile implant with foam and lead protection of the testicles
(Fig. 10.3 in text).

6.8 **Radiobiology**

The initial dose-rate at the 145 Gy reference isodose is 7.7 cGy/hr. The dose-rate falls with a half-life of 59.4 days and the dose rapidly accumulates over the first few weeks of implantation and then tails off.

The dose-rate for palladium is higher but falls off more quickly. There is a suggestion that side effects therefore reach a peak sooner and settle a little sooner than for Iodine 125.

6.9 **Post-implant dosimetry**

When a patient fails after prostate brachytherapy, it is important to know whether the failure was due to poor selection or to poor technique. In order to evaluate technique it is necessary to establish quality indices which can be evaluated post implantation and then to see how these correlate to either recurrence or the risk of side effects.

The main determinant of successful treatment is achieving the right dose to the right volume. The V100 is the percentage of the target volume that has received the prescription dose. This should be greater than 90%.

The D90 is the dose which has been received by 90% of the target volume. This should be at least 90% of the prescribed dose of 145 Gy.

The V150 is the volume which receives greater than 150% of the prescribed dose. This gives an indication of the homogeneity of dose distribution and may be related to the risk of side effects.

The maximum rectal dose is related to the risk of proctitis. If kept less than 200 Gy the risk is small.

The maximum urethral dose should be kept to less than 150% of the treatment dose, i.e. 220 Gy.

6.10 **Patient care during implant**

6.10.1 **Pre-implantation**

The patient is usually given an enema before implantation to ensure a good field of vision for the transrectal ultrasound.

Patients on aspirin should stop it at least a week before treatment to reduce the risk of bleeding.

6.10.2 **Post-implantation care**

Immediately post-implant it is important to image the implant as shown in

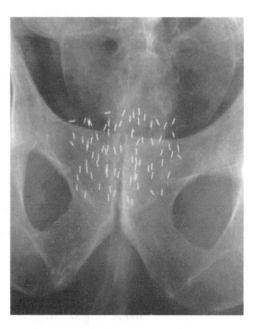

Fig. 6.4 Radiograph of iodine 125 seeds in the prostate.

Fig. 6.4, and count the seeds to ensure that none have been misplaced during the procedure. Perineal bruising after the implant can be helped by application of an ice pack.

Patients are asked to pass urine through a sieve during their time in hospital to catch any seeds which may pass into the bladder or urethra. This is uncommon if stranded seeds are used.

Irritative and obstructive urinary symptoms are very common and can be helped considerably by alpha blocking drugs. These are usually started the day before operation and may need to be taken for several months until symptoms settle.

Pain on micturition is helped by anti-inflammatory drugs.

Post-implant infection is managed by prophylactic antibiotics, Ciprofloxacin 500 mg twice a day for 7 to 10 days is commonly used.

6.10.3 Outcome

Local control is difficult to assess either digitally or by imaging. Survival is a relatively poor measure of the effectiveness of treatment because 30% or more of elderly men die of unrelated causes. The best outcome measure is the PSA relapse free survival. Patients are considered to have failed brachytherapy if they have three consecutive rises in PSA with at least three months between each measurement. About 25 to 30% of patients show the phenomenon of benign PSA bounce where one or two rises in PSA may occur followed by a slow fall. The cause is unclear but it does not have any significant impact on outcome. It may take several years for the PSA to achieve its nadir and it seems that those in whom the PSA falls slowly do a little better than those in whom it falls quickly. There is still some controversy about a nadir level which may define success but if the PSA does not fall to less than 1.0 µg/L, the majority of patients eventually suffer recurence.

For good prognosis patients (PSA less than 10, Gleason score 6 or less) PSA relapse free survival at 5 years is 80 to 90%.

For intermediate patients (either Gleason score greater than 6 or PSA greater than 10) five year PSA relapse free survival is 60 to 70%. For poor prognosis patients (PSA greater than 10, Gleason score 7 or greater) PSA relapse free survival at five years is 40 to 50%.

6.11 Side effects and complications

The main risk of side effects is to urinary, bowel, and sexual function.

6.11.1 Urinary side effects

Nearly all patients get temporary urethritis. This may be acute for 2 to 3 weeks but urinary irritability may continue for several months. Approximately 50% of patients still have mild or moderate symptoms at six months, but by one year only 2 or 3% of patients still have urethral symptoms.

Overall approximately 15% of patients may develop acute urinary retention usually in the first two weeks after implantation. The risk is higher for patients with large prostates and high prostate symptom scores.

Acute retention can mostly be managed by urethral catheterization but a suprapubic catheter is occasionally needed. Some patients can be managed by intermittent self-catheterization. Most patients are free of catheter after three or four weeks, but in approximately 5% it may be necessary to leave a catheter in for three months or more.

TURP to relieve obstructive symptoms should be avoided if possible because of a risk of producing urethral necrosis and incontinence. If necessary it should be delayed for a year if possible and then done with the minimum procedure necessary to restore the outflow.

6.11.2 Rectal complications

Approximately 5% of patients may develop proctitis, intermittent rectal bleeding, and some discomfort. The vast majority settles by one year. The risk of fistula reported in the literature from brachytherapy alone is 0.1 to 0.2%.

6.11.3 **Sexual function**

Many patients referred for brachytherapy already have erectile dysfunction. For those who do not, 30 to 40% become impotent in the year or two following brachytherapy. The risk is less in young men who are fully potent and greater in older men whose potency is already waning. In the vast majority of cases erectile dysfunction is resolved by Viagra.

6.12 **Radioprotection**

The range of the radiation released from iodine seeds is very short and the radiation hazard to others is negligible. Standard guidelines however recommend that close contact (i.e. <1 m) with young children and pregnant women is avoided for two months after the implant. There is also uncertainty with regard to the hazard from cremation in the early period post-implant and guidelines suggest avoiding cremation for up to two years post-implant.

6.13 **Points and pitfalls**

Perhaps more than any other form of brachytherapy, permanent seed implantation for prostate cancer requires team working. This requires collaboration between the oncologist, urologist, and radiologist with expertise in ultrasound plus a brachytherapy physicist and a dedicated theatre and ward team. There is a significant learning curve and attendance at a teaching course and some mentoring is essential to get started.

6.14 **High dose-rate afterloading brachytherapy for localized prostate cancer**

Whilst the most common form of brachytherapy for prostate cancer is that of permanent seed implantation, high dose-rate afterloading techniques are an alternative. The principal differences are as follows:

1. There is no pre-or per implant dosimetry. Catheters are placed in situ and post-implant dosimetry undertaken.

2. There is greater flexibility in volume and dosimetry which is defined by catheter position and dwell times.

3. HDR brachytherapy enables large fraction sizes to be delivered to the prostate volume. If prostate cancer has a low alpha-beta ratio, as has been postulated, this may have distinct biological advantages.

6.14.1 Indications

HDR brachytherapy for prostate cancer is predominantly used as a boost after external beam treatment. HDR monotherapy is currently being explored and may provide an alternative to seed monotherapy.

6.14.2 Implant procedure

- The patient preparation and positioning are identical to those for seed brachytherapy. The patient is given a general or spinal anaesthetic and the implant is performed with the patient in the lithotomy position. Transrectal ultrasound guidance is used.

- An indwelling urinary catheter should be placed and will remain throughout the period of treatment.

- A means of fixation to the skin is required which may be in the form of a rigid template used in conjunction with the ultrasound or a flexible template used in parallel to the ultrasound template.

- High dose-rate afterloading catheters are placed within the prostate gland using transrectal ultrasound control in exactly the same way as seed needles are inserted using the co-ordinates of the template to define their position. Applicators may be hollow steel needles or flexible plastic applicators.

- Catheter insertion may follow one of two conventions:
  - (a) Peripheral weighting with catheters predominantly around the periphery of the gland with a smaller number centrally.
  - (b) Individualized needle placement for localized boosting.

- At completion of the implant the catheters or applicators must be fixed to the perineum. A rigid template left in situ sutured to the skin is one method. A flexible template fixed to the skin with adhesive using flexible plastic catheters is an alternative method. This is shown in Fig. 6.5.

- After completion of the implant CT imaging is undertaken and the implant reconstructed on a three dimensional planning system.

- The CTV and PTV are defined on the CT images. Typically the CTV will be the prostate defined by its capsule and the PTV will be this volume grown by 2–3 mm in all dimensions. The seminal vesicles may also be included in the CTV when indicated, catheters being introduced to include these as shown in Fig. 6.6.

- The dose distribution is defined by the dwell positions in the HDR dosimetry programme. Where the whole gland is to be treated, a modified peripheral loading approach will be used weighting the peripheral catheters and the proximal and distal ends of the catheters.

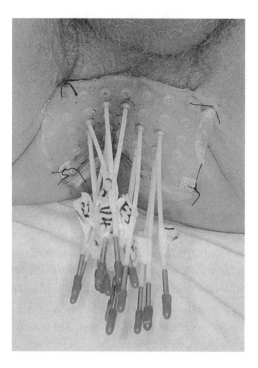

Fig. 6.5 HDR prostate implant using flexible implant system *in situ* (see colour Plate 7).

- Treatment delivery follows in the HDR afterloading room.
- Quality assurance is important to ensure that catheters do not move between treatments. This is particularly the case where 2, 3, or 4 fractions may be given with the same implant. The distal catheter length from the template to the connecting hub should be measured for each treatment to ensure there has been no movement at the skin, and ideally, a limited CT series undertaken before each fraction to evaluate the internal movement of the prostate gland. Dwell positions or catheter positions may need adjustment for each fraction.
- On completion of treatment the catheters and template are removed.
- The indwelling urinary catheter will also be removed.

6.14.3 Dosimetry

Within the PTV will be the urethra, which is one of the limiting organs at risk. The other structure to be considered is the anterior rectal wall. Dosimetry will be designed to have a relative cold spot around the urethra and to limit the anterior rectal wall dose. This is shown in Fig. 6.7.

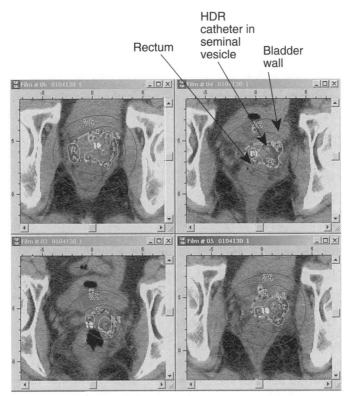

Fig. 6.6 CT planning scan of HDR implant showing implant extended to include seminal vesicles (see colour Plate 8).

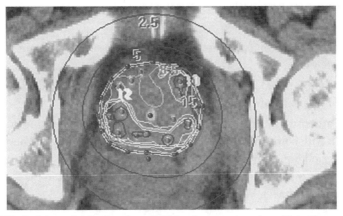

Fig. 6.7 Typical dose distribution from HDR implant (see in colour, Plate 9).

6.14.4 **Prescription**

A range of prescriptions are used in HDR afterloading brachytherapy reflecting both the relative infancy of this treatment technique and also the variations in prescription definition. Examples include the following:

(a) After external beam treatment delivering the equivalent of 45 Gy in 25 fractions

6 Gy × 3

8.5 Gy × 2

11 Gy × 2

15 Gy × 2

When comparing implant prescriptions from different centres it is important to consider the total distribution and in particular, the dose to organs at risk, especially the anterior rectal wall alongside the defined tumour dose.

(b) HDR monotherapy programmes are currently evaluating this technique. Prescriptions used include 34 to 36 Gy in 4 fractions.

6.14.5 **Outcome of treatment**

There is no randomized evaluation of HDR brachytherapy compared with conventional optimized high dose external beam conformal radiotherapy. In theory, it offers an alternative means of delivering high dose conformal radiation therapy to the prostate gland. Phase II studies suggest that the local control rates are no worse than those with optimal external beam programmes.

6.14.6 **Complications**

Complications are those of radical radiotherapy to the prostate gland. Particular considerations where brachytherapy is used are:

1. *Acute toxicity*

 • Perineal bruising and soreness is inevitable, usually self-limiting within a few days.

 • Dysuria may occur for a few days after treatment, but tends to be less prominent than with seed implantation.

 • Particularly when given as a boost with external beam treatment, proctitis with bowel frequency and urgency will persist during and beyond the period of brachytherapy. The constipating process for treatment delivery may also make bowel symptoms more prominent in the few weeks after treatment.

2. *Late effects*

Urethral stenosis has been reported in up to 7% of patients. No other specific complications distinct from those expected with external beam treatment (e.g. late radiation proctitis and erectile impotence) have been identified.

Further reading

Seed brachytherapy

Beyer DC, Priestley JB (1997). Biochemical disease free survival following iodine 125 prostate implantation. *International Journal of Radiation Oncology, Biology and Physics* 7:1035–9.

Gelblum DY, Potters L, Ashley R, *et al.* (1999). Urinary morbidity following ultrasound guided transperineal prostate seed implantation. *International Journal of Radiation Oncology, Biology and Physics* 45:59–67.

Stock RG, Stone NN, Tabert A, *et al.* (1998). A dose response study for iodine 125 prostate implants. *International Journal of Radiation Oncology, Biology and Physics* 41:101–8.

Zelefsky MJ, Hollister T, Raben A, *et al.* (2000). 5 year biochemical outcome and toxicity with transperineal CT planned permanent I125 prostate implantation for patients with localized prostate cancer. *International Journal of Radiation Oncology, Biology and Physics* 47:1261–6.

Ash D, Flynn A, Batterman J, *et al.* (2000). ESTRO/EAU/EORTC recommendations on permanent seed implantation for localized prostate cancer. *Radiotherapy & Oncology* 57:315–21.

Potters L (2003). Permanent prostate brachytherapy in men with clinically localized prostate cancer. *Clinical Oncology* 15:301–15.

High Dose-Rate afterloading

Hoskin PJ (2001). High dose-rate afterloading brachytherapy for prostate cancer. In: Joslin C, Flynn A, Hall E (ed.). *Principles and practice of brachytherapy using afterloading systems*, pp 257–65. London:Arnold.

Hoskin PJ (2000). High dose-rate boost treatment in radical radiotherapy for prostate cancer. *Radiotherapy and Oncology* 57:285–88.

Galalae RM, Martinez A, Mate T *et al.* (2004). Long term outcome by risk factors using conformal high-dose-rate brachytherapy (HDR-BT) boost with or without neoadjuvant androgen suppression for localized prostate cancer. *International Journal of Radiation Oncology, Biology and Physics* 58:1048–55.

Chapter 7

Endolumenal brachytherapy: bronchus and oesophagus

Peter Hoskin

7.1 Introduction

With the advent of high dose-rate (HDR) afterloading systems using fine catheters of 2 mm diameter, it is now possible to pass high dose-rate afterloading sources readily into the bronchus or oesophagus. The techniques are well defined but their place in the management of these tumours where external beam has a well established role remains uncertain. As with all types of brachytherapy, HDR afterloading offers the opportunity to deliver a high localized dose of radiation to a defined area within the bronchus or oesophagus.

7.1.1 Indications

1. In combination with external beam treatment as a radical course of radiotherapy.
2. Palliative treatment as the sole means of delivering radiation.

7.1.2 Advantages

Catheters can be placed intraluminally in the bronchus or oesophagus with relative ease requiring moderate sedation at bronchoscopy or the passage of a nasogastric tube. Most treatments are delivered as single doses and therefore a day case or one night stay in hospital is sufficient.

7.1.3 Disadvantages

1. Passage of an endobronchial catheter at bronchoscopy is a skilled procedure requiring a well trained team.
2. Dose penetration is limited, the typical prescription point being 1cm from the source which in many bronchial tumours may not be sufficient to cover the entire tumour.

3. Unless some attempts are made to centre the catheter within the lumen of the bronchus or oesophagus it is usual that the dose will be offset to one side or the other.

4. Tumours close to the cardia and extending into the fundus of the stomach may not be well covered by a central catheter.

5. Tumours of the carina or bilateral bronchial tumours require passage of two catheters and more complex dosimetry.

7.2 Implant procedure

7.2.1 Bronchus

Fibreoptic bronchoscopy is performed under light sedation or if preferred under general anaesthesia. Specialist anaesthetic support will be required to allow passage of the bronchoscope through the nasal cavity and maintenance of airway during the procedure. The tumour is identified and the extent defined.

1. The HDR afterloading catheter is passed down the suction channel of the bronchoscope and under direct vision passed alongside and beyond the tumour bearing area of bronchus as shown in Fig. 7.1.

2. The bronchoscope is withdrawn over the HDR catheter leaving it *in situ*. This is best performed by two operators, one controlling the catheter and the other withdrawing the bronchoscope.

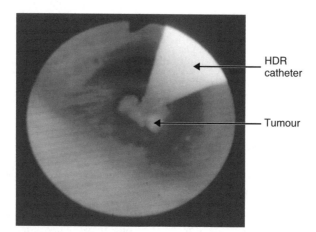

Fig. 7.1 View of endobronchial tumour and HDR catheter passed beyond it through bronchoscope (see in clolour, Plate 10).

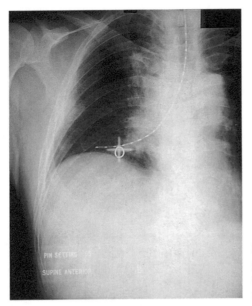

Fig. 7.2 X-ray to show position of HDR catheter within bronchus after removal of bronchoscope.

3. The catheter is retained by careful fixation at the nose. Note that the catheter should be introduced through the nose and not the mouth where damage and distortion from the teeth can occur.

4. Following recovery from sedation or anaesthetic AP and lateral X-rays are taken, as shown in Fig. 7.2. If required, CT scans can also be used to supplement this information.

5. The treatment length is defined on the X-ray or CT relating back to fixed anatomical structures such as the carina, the bronchoscopic appearances and description and knowledge of catheter position. The distance from the catheter end dwell position, the origin, must also be defined to allow the treatment to be delivered at the correct position along the catheter length.

6. Dwell positions are defined along the treatment length to deliver a prescription dose at 1 cm from the source. Note that if the end dwell positions are not weighted then there will be tapering of the dose at each end of the defined treatment length. A typical distribution is shown in Fig. 7.3.

7. Treatment is delivered by attachment of the HDR catheter to the afterloading machine and passage of the source along the defined length.

8. Following treatment the catheter is removed. Sedation is not usually required for this and the patient then returns to the ward to recover and later be discharged.

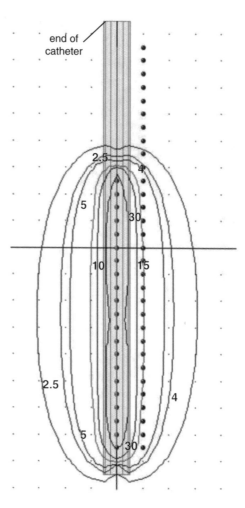

Fig. 7.3 Dose distribution for HDR catheter as used in endobronchial or endo-oesophageal treatment showing unweighted distribution with 5 cm origin.

7.2.2 **Oesophagus**

1. Prior to insertion, an HDR afterloading catheter is passed into a gauge 16 nasogastric tube ensuring it passes to the end of the tube and is fixed by taping at the open end as shown in Fig. 7.4.

2. The nasogastric tube is passed into the stomach through the area of known tumour.

3. AP and lateral X-rays are taken and if required CT scans to supplement this information.

4. Where plain X-rays are taken, it may be useful to repeat the films following the initial X-ray with the patient having swallowed oral contrast to define the level of the tumour, as shown in Fig. 7.5.

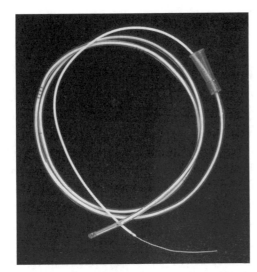

Fig. 7.4 HDR catheter preloaded into nasogastric tube in preparation for endo-oesophageal treatment.

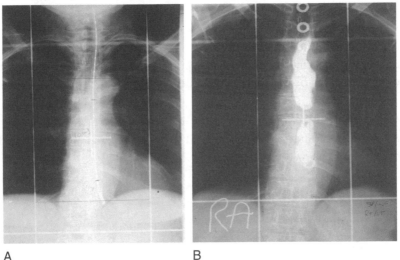

A   B

Fig. 7.5 X-ray to show nasogastric tube and HDR catheter in position (A) and after administration of oral contrast to identify site of tumour stricture (B).

5. The treatment length can then be defined on the plain film.

6. The dwell times required for the prescription dose are calculated. Conventionally, this will be defined at 1 cm from the source.

7. Treatment is delivered by insertion of the HDR catheter into the afterloading machine.

8. Following treatment, the nasogastric tube can be withdrawn and the patient returns to the ward for recovery and discharge.

7.3 Volume definition

The position of the catheter in the lumen of the oesophagus or bronchus will depend upon its calibre. Typically, where there is tumour, the lumen will be narrowed and the catheter will lie approximately centrally, particularly in the oesophagus where it is already lying within a nasogastric tube. In a large lumen, however, such as the trachea or main bronchus, centring catheters can be employed with inflatable ears, which centre the catheter once in position.

7.3.1 Bronchus

Typically, a cylinder of 1 cm radius will be treated. The length of the cylinder will be defined by the direct vision description of the tumour. A margin of 1 to 2 cm proximally and distally is introduced, noting that there will be tapering of doses at the distal ends of the cylinder unless the dwell times are weighted at each end. This is not usual practice.

Tumours at the carina present a particular challenge to endobronchial brachytherapy. A decision may be made to only treat the dominant area of the tumour extending into the adjacent main bronchus or more complex techniques using a second catheter have been described.

7.3.2 Oesophagus

A 1 cm radius cylinder is typically treated. A 2 to 3 cm margin proximally and distally will again be used.

7.4 Dose prescription

7.4.1 Bronchus

(a) Radical treatment: This is not standard practice, but case series have been reported after external beam doses of 40 to 60 Gy. Additional boost doses of 6 to 12 Gy in 1 to 2 fractions are given.

(b) Palliation is typically given in a single dose of 10 to 15 Gy defined at 1 cm.

7.4.2 Oesophagus

(a) Radical treatment has been described delivering 40 Gy in 2 Gy fractions followed by 1 or 2 fractions of 10 Gy at 1 cm. CHART (40.5 Gy in 27 fractions in 9 days) has been followed by a single dose of 15 Gy.

(b) Palliation is typically delivered with a dose of 15 Gy at 1 cm. A randomized trial has compared 18 Gy in 3 fractions with 16 Gy in 2 fractions and no difference in symptom control, survival, or complications were seen.

7.5 Clinical results

7.5.1 Bronchus

Case series attest to the efficacy of endolumenal brachytherapy in the management of symptoms from advanced inoperable non-small cell lung cancer, in particular cough, haemoptysis, and dyspnoea. One randomized controlled trial, however, has shown no compelling advantage of endolumenal brachytherapy over an external beam course of 32 Gy in 8 fractions, the brachytherapy being superior in terms of fatigue and nausea, but inferior for the management of chest pain and dyspnoea.

In conclusion therefore, there is no good evidence to support the use of brachytherapy as primary palliative treatment, but in circumstances where there has been previous radiation, re-treatment with brachytherapy may be considered. It is equivalent to external beam treatment for palliation of cough and haemoptysis and may be considered as an appropriate alternative where available.

7.5.2 Oesophagus

One small trial suggests improved local control when a brachytherapy boost was given after 60 Gy external beam radiotherapy compared to an external beam boost, particularly in tumours <5 cm in length. There is no additional evidence to support its use in radical treatment or in combination with chemoradiation where an increase in complications has been reported in one phase II study.

Palliation case series suggest that brachytherapy is effective in the management of progressive dysphagia from carcinoma of the oesophagus but comparative data against other manoeuvres including intubation and stent insertion, laser or cryotherapy, and external beam treatment are not available. All treatments seem effective, surgical options are more immediately effective. Brachytherapy may be useful following insertion of a stent, laser therapy, or cryotherapy.

7.6 Complications

7.6.1 Bronchus

Acute complications include local discomfort and an increased cough for several days after the procedure, often with haemoptysis because of local bruising. This settles spontaneously.

Late complications are few. An increased incidence of intrabronchial haemorrhage was described in individual case series but further analysis suggests that this is no more common than after other treatments for non-small cell lung cancer.

7.6.2 Oesophagus

Acute side effects with oesophagitis and dysphagia is to be expected, self-limiting over a period of 10 to 14 days.

Late side effects include oesophageal stricture, particularly when given as a boost after external beam treatment. When given as sole palliative treatment, late complications are rare, although long term tumour control is also not expected and further oesophageal symptoms may occur.

Further reading

Bronchus

Gollins SW, Burt PA, Barber PV, Stout R (1994). High dose-rate intraluminal radiotherapy for carcinoma of the bronchus: outcome of treatment of 406 patients. *Radiotherapy and Oncology* **33**:31–40.

Gollins SW , Burt PA, Barber PV, Stout R (1996). Long Term Survival and Symptom Palliation in Small Primary Bronchial Carcinomas Following Treatment with Intraluminal Radiotherapy Alone. *Clinical Oncology* **8**:239–46.

Gollins SW, Ryder WDJ, Burt PA, Barber PV, Stout R (1996). Massive haemoptysis death and other morbidity associated with high dose-rate intraluminal radiotherapy for carcinoma of the bronchus. *Radiotherapy and Oncology* **39**:105–16.

Speiser BL. Endobronchial brachytherapy in the treatment of lung cancer. In: Joslin CA, Flynn A, Hall E (ed.). *Principles and practice of brachytherapy using afterloading systems,* pp 225–42. London: Arnold.

Stout R, Barber P, Burt P, Hopwood P, Swindell R, Hodgetts J, Lomax L (2000). Clinical and quality of life outcomes in the first United Kingdom randomised trial of endobronchial brachytherapy (intraluminal radiotherapy) vs. external beam radiotherapy in the palliative treatment of inoperable non-small cell lung cancer. *Radiotherapy and Oncology* **56**:323–7.

Van Limbergen E, Potter R (2002). Bronchus cancer. In: Gerbaulet A, Potter R, Mazeron JJ, Meertens H, Evan Limbergen (eds.). *The GEC ESTRO handbook of brachytherapy,* pp 545–60. Brussels: ESTRO.

Oesophagus

Brewster AE, Davidson SE, Makin WP, Stout R, Burt PA (1995). Intraluminal brachytherapy using the high dose-rate microselectron in the palliation of carcinoma of the oesophagus. *Clinical Oncology* **7**:102–5.

Flores AD (2001). Brachytherapy in cancer of the oesophagus. In: Joslin CA, Flynn A, Hall E (ed.). *Principles and practice of brachytherapy using afterloading systems,* pp 243–56. London: Arnold.

Gaspar LE, Nag S, Herskovic A, Mantravadi R, Speiser B and the Clinical Research Committee, American Brachytherapy Society, Philadelphia, PA (1997). American Brachytherapy Society (ABS) Consensus Guidelines for Brachytherapy of esophageal cancer. *International Journal of Radiation Oncology, Biology and Physics* **38**:127–32.

Gaspar LE, Winter K, Kocha WI, Coia LR, Herskovic A, Graham M (2000). A Phase I/II Study of External Beam Radiation, Brachytherapy, and Concurrent Chemotherapy for Patients with Localized Carcinoma of the Esophagus (Radiation Therapy Oncology Group Study 9207). *Cancer* **88**:988–95.

Okawa T, Dokiya T, Nishio M, Hishikawa Y, Morita K and the Japanese Society of Therapeutic Radiology and Oncology (JASTRO) Study group (1999). Multi-institutional randomized trial of external radiotherapy with and without intraluminal brachytherapy for esophageal cancer in Japan. *International Journal of Radiation Oncology, Biology and Physics* **45**:623–28.

Potter R, van Limbergen E (2002). Oesophageal cancer. In: Gerbaulet A, Potter R, Mazeron JJ, Meertens H, Evan Limbergen (eds.). *The GEC ESTRO handbook of brachytherapy*, pp 505–14. Brussels: ESTRO.

Sur RK, Levin CV, Donde B, Sharma V, Miszczyk L, Nag S (2002). Prospective randomised trial of HDR brachytherapy as a sole modality in palliation of advanced esophageal carcinoma-an International Atomic Energy Agency study. *International Journal of Radiation Oncology, Biology and Physics* **53**:127–33.

Chapter 8

Perineal implants: anal canal, vagina, and vulva

Peter Hoskin

8.1 Introduction

Perineal tumours are typically squamous cell carcinomas readily accessible to interstitial or intralumenal brachytherapy. Brachytherapy should be considered in the radical radiotherapeutic management of these sites and it may also have an important role for the palliation of locally advanced or recurrent disease.

8.2 Indications

8.2.1 Anal canal

1. brachytherapy alone for small T1 tumours;
2. as a boost after radical chemoradiation for larger tumours;
3. as a palliative treatment for recurrent or locally advanced tumours.

8.2.2 Vaginal tumour

1. brachytherapy alone or in combination with external beam treatment for stage I tumours;
2. as a boost after external beam treatment for stage II and III tumours;
3. as a palliative local treatment for stage IV and recurrent tumours.

8.2.3 Vulva

1. as primary treatment for small T1 tumours;
2. as a boost after external after external beam treatment for more locally advanced tumours;
3. as a palliative treatment for locally advanced or recurrent tumours.

8.3 **Implant procedure**

Both low dose-rate iridium wire interstitial implants and high dose-rate after-loading implant techniques can be used in this setting. For both, the implant procedure is similar taking account of the different applicators required but low dose-rate requires patient isolation and involves greater staff exposure.

For selected superficial tumours of the vagina or anal canal, intracavitary treatment may be appropriate. For vaginal tumours, the technique will use vaginal applicators similar to those described in Chapter X but since, in general tumours will not be in the fornices, a stump applicator is not specifically required. Doses can be prescribed up to 1 cm from the surface with tolerable surface doses using a 2 cm diameter applicator. This technique therefore is limited to tumours where the PTV will not be more than 1 cm from the surface. In practice, this includes tumours which have been completely excised by excision biopsy or flat superficial tumours. Even in these settings there is a significant risk of microscopic nodal disease particularly in poorly differentiated tumours for which some external beam treatment should be considered. In the palliative setting, however, intralumenal treatment is simpler and quicker and may therefore provide an important alternative to a full interstitial implant.

8.4 **Interstitial implant technique**

8.4.1 **Tumour localization**

Tumour localization for all these sites will be predominantly based upon clinical examination supplemented by CT or MR scanning to determine the extent of soft tissue infiltration and invasion of the surrounding structures.

Vaginal or rectal ultrasound may also be valuable and can be used during the implant to identify catheter position.

8.4.2 **Applicators**

Perineal implants will be best performed using some form of perineal template. This has two functions:

1. To direct applicator position and ensure even spacing of sources

2. To provide fixation of the source-carrying applicators

There are a number of commercially available applicators and many departments have their own in-house design. Some of these are illustrated in Fig. 8.1. The basic principles of these applicators are:

- a plate resting against the perineal surface with holes evenly spaced to carry the applicators;

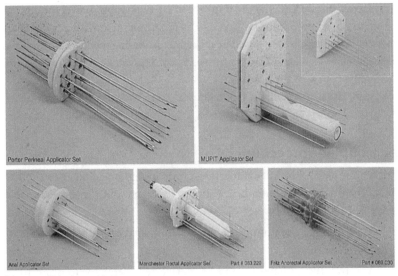

Fig. 8.1 Perineal templates.

- allowance for a urinary catheter;
- an obdurator which, depending upon the site of implant, may be a rectal or vaginal obdurator. This will have two functions:
  1. To displace the contralateral wall of the vagina or ano-rectal canal away from the tumour and hence the implant, where less than the full circumference is to be implanted. This can be achieved equally well using vaginal gauze packing.
  2. In the case of ano-rectal implants to provide a conduit for the passage of flatus and fluid.

The implant will be carried out with hollow applicators, which will pass through the template holes. Typically these will be either hollow needles, or in the case of HDR, hollow plastic tubes may be used.

There are alternative approaches. Interstitial iridium hairpins can be used for a low dose-rate implant to the anal canal. They may provide a means of implantation without the use of a template, but have the disadvantage of a less rigid set up and without an obdurator, unless gauze packing is used, less control of the contralateral ano-rectal canal is achieved.

8.4.3 Patient preparation

Depending upon the dose prescription the implant is likely to be in position for several days. It is important that the implanted area is free from contamination with faeces and that during the implant, as far as possible, bowel motions are avoided. Bowel preparation, therefore, is recommended with a laxative and enema the night before, followed by regular constipating medication, for example codeine phosphate 30 mg four times a day. This will also provide analgesia during the course of the implant.

Urinary drainage should be maintained by an indwelling catheter throughout the procedure.

8.4.4 Implant procedure

Perineal implant should be undertaken only under a general anaesthetic or spinal anaesthetic. The procedure is similar for each of the three sites as shown in Fig. 8.2 and detailed below:

- The position of the tumour should be confirmed on clinical examination under anaesthetic.
- An indwelling catheter should be placed per urethrum to drain urine, taking account of its position within any template system used.
- The template should be placed in position. Some templates require sutures to the skin to anchor the perineal plate; an alternative is to use a flexible template which can remain fixed to the skin by adhesive.
- Once in position, the applicators are simply passed through the appropriate holes of the template and guided along the wall of the vagina or anal canal or in the case of a vulval implant through the tumour.
- It may be difficult to judge the appropriate depth to which the applicator should be placed. This may be pre-determined by measurements on clinical examination and scans to determine the depth of the PTV and the applicator is then inserted to that depth, bearing in mind that if a Paris dosimetry system is used, the applicator should be over-inserted beyond the edge of the PTV.
- Where the template permits access, the position of the applicator can be further verified by digital examination or by an ultrasound probe measurement.
- Applicator insertion will typically occur with the patient in the lithotomy position. It should be borne in mind that when the legs are brought down into the straight position, (the position in which treatment would usually be given), there is a divergence of the applicator ends and insertion in the lithotomy position should therefore aim for a small degree of convergence which could then be corrected to achieve parallel source positions when the legs are straight.

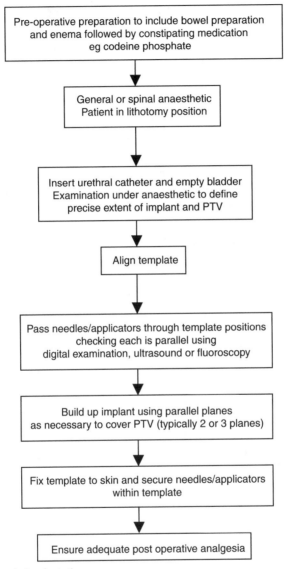

Fig. 8.2 Steps in implantation.

- Once the applicators are in position, fixation should be ensured either by a locking mechanism within the template, individual suturing to the skin surface, or use of an adhesive.
- At completion of the implant, the length of catheter protruding beyond the template should be measured and carefully recorded to aid dosimetry,

definition of dwell positions or wire length, and for subsequent quality assurance to ensure that the implant has not moved.

• Padding is placed around the implant to prevent trauma to the thighs and the patient is then transferred for CT scanning and treatment delivery.

8.5 Dosimetry

Where steel needles have been used orthogonal X-rays may be taken. However, in current practice, a CT scan sequence through the implant is preferred. The CT scan could then be transferred into the planning system, the PTV defined by the clinician and the loading of each applicator defined by the physicist.

If LDR is being used then there is already a pre-implant calculation of the expected PTV source numbers and lengths so that iridium wire of an appropriate activity and length could be ordered. The wires are then cut and prepared for afterloading in a plastic tube.

For a HDR implant the dwell times and positions together with the source origin will be defined using the computer algorithm. An optimization programme will enable the best solution for an even distribution typically employing uneven dwell times weighted towards the periphery of the implant. An example is shown in Fig. 8.3.

8.6 Treatment delivery

LDR implants will require loading using long-handled forceps and timing the implant loading to enable removal of the implant within the normal working day. This should be undertaken in the protected room with full radioprotection precautions according to the activity of the implant.

Daily quality assurance checks should be undertaken with measurement of the applicator ends from the template to ensure no movement has taken place. HDR afterloading implants will be treated in the HDR treatment room. Before each fraction the catheter lengths to the template should be measured to ensure no movement has taken place. Some advocate CT scanning also to assess internal changes, in particular, tissue oedema which may alter the position of the PTV along the length of the catheter. This may be corrected for by an adjustment of the origin of the source dwell times.

8.7 Dose prescription

8.7.1 Low dose-rate

Sole treatment   60 Gy at 0.5 Gy per hour

Boost   After external radiotherapy or chemoradiation delivering 45 to 50 Gy should be to a dose of 20 to 25 Gy at 0.5 Gy per hour

(a)

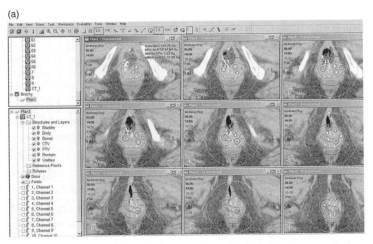

(b) Dwell times (secs at 5 mm intervals for 10 catheter perineal implant)

Catheter	Position (5 mm steps)										
	1	2	3	4	5	6	7	8	9	10	11
1	2.8	3.0	1.5	2.8	2.7	1.8	1.9				
2	9.7	9.7	9.6	7.0	2.8	4.5	6.3				
3	5.1	5.7	2.9	1.5	0.5	0.2	0.6	2.1	3.1		
4	0	0	9.1	6.7	4.8	1.2	1.1	0.9	1.5	2.1	
5	0.3	1.8	2.0	4.8	3.0	1.0	4.5	0			
6	5.5	4.2	3.5	0.2	0.4	0.3	1.2	1.5	1.9	1.0	0.7
7	2.8	1.8	1.8	1.4	3.1	5.6	5.8	6.1	6.2	6.9	5.5
8	0	3.7	4.8	3.8	3.8	2.7	0	0.2	2.8	0.1	2.7
9	0	0	2.6	0	0	0.9	0	0	0	0	0.1
10	7.9	5.3	1.3	0.1	0	0	0	0	0	0	3.0

Fig. 8.3 CT based implant dosimetry (a) and catheter dwell times (b) (see (a) in colour, Plate 11).

8.7.2 **HDR afterloading**

Sole treatment   30 to 36 Gy in 5 to 6 fractions
As a boost after external beam treatment delivering 45 to 50 Gy should be 16.5 Gy in 3 fractions

Palliative treatment will be individualized. An LDR implant delivering 20 to 30 Gy will give good palliation as will an HDR implant delivering a single dose of 10 Gy or 18 to 24 Gy in 3 to 4 fractions.

8.8 **Clinical results**

8.8.1 **Anal cancer**

After 40 to 50 Gy external beam with or without concomitant chemotherapy, the results of LDR implantation give a high complete response rate and local control rate of around 90% and 80%, respectively, for selected patients. There has been some concern that brachytherapy results have a higher incidence of necrotic ulcers compared to external beam techniques. This is almost certainly operator and technique dependent.

8.8.2 **Vaginal cancer**

High rates of local control with localized vaginal cancer can be achieved with radical radiotherapy. For stage I disease, 80% local control rates and disease free survival are reported. Pelvic control is lower once there is spread beyond the vaginal walls; however, there is some evidence to suggest that the addition of brachytherapy to external beam radiotherapy increases local control compared to external beam alone in both stage II and stage III disease. For stage II disease with external beam alone, a 50% pelvic control rate is increased to 80% when brachytherapy is included. For stage III, a 30% pelvic control rate is increased to over 55% when brachytherapy is included. Complications include proctitis, bladder neck and urethral strictures, and a risk of fistulae, particularly where there has been extensive tumour breakdown. In stage III and IV disease, 10% of patients are reported to have either a recto-vaginal or vesico-vaginal fistula.

8.8.3 **Vulval cancer**

Primary radiotherapy is generally reserved for patients unfit or unwilling to undergo surgery or for those with advanced inoperable disease. There is no prospective study comparing radiotherapy with surgery in this disease. Overall, local tumour control rates of around 50% are reported in various series. Complications include vulvitis, urethral stricture, fistulae, and necrosis; the incidence in published series varies from zero to 50%.

Further reading

Mock U, Kucera H, Fellner C, Knocke TH, Potter R (2003). High-dose-rate (HDR) brachytherapy in the treatment of primary vaginal carcinoma: Long-term results and side effects. *International Journal of Radiation Oncology, Biology and Physics* **56**:950–7.

Perez CA, Camel MH, Galakatos AE, Grigsby PW, Kuske RR, Buchsbaum G , Hederman MA (1988). Definitive irradiation in carcinoma of the vagina: Long-term evaluation of results. *International Journal of Radiation Oncology, Biology and Physics* **15**:1283–90.

Puthawala AA, Nisar Syed AM (2001). Interstitial brachytherapy in the treatment of carcinoma of the anorectum. In: Joslin CA, Flynn A, Hall E (ed.). *Principles and practice of brachytherapy using afterloading systems*, pp 387–92. London: Arnold.

Sandhu APS, Symonds RP, Robertson AG, Reed NS, McNee SG, Paul J (1998). Interstitial iridium-192 implantation combined with external radiotherapy in anal cancer: Ten years experience. *International Journal of Radiation Oncology, Biology and Physics* **40**:575–81.

Chapter 9

Breast brachytherapy

Peter Hoskin

9.1 Introduction

Breast irradiation forms a major part of the activity in any radiotherapy centre. The vast majority of patients receive external beam treatment and only a selected minority are considered for brachytherapy. This, in part, reflects the availability of expertise and resources to deliver brachytherapy and, in part, the efficiency and efficacy of breast boost treatments with electron beams. There are however specific indications where brachytherapy should be considered, in particular, where a high dose of radiation is required to an underlying breast tumour, which might exceed skin tolerance. The advantages for a brachytherapy breast boost over external beam radiotherapy, therefore, lie in the ability to deliver a high tumour dose with skin sparing to a tumour within the breast, and also reduce dose to underlying structures, in particular, the ribs and lungs. In this setting, for example, where there is residual tumour for which local excision is not to be undertaken, better cosmesis may well be achieved with brachytherapy, than attempting to deliver a high dose through the skin with external beam treatment.

9.2 Indications

1. Routine post-operative boost treatment following whole breast external beam radiotherapy.
2. Locally advanced tumour following neo-adjuvant chemotherapy and external beam radiotherapy.
3. Palliation of advanced disease.

9.3 Advantages

Enables a high tumour dose with the option to design into the implant dosimetry, skin sparing and lung and rib sparing.

9.4 **Disadvantages**

1. Can only treat a limited volume and becomes impractical for tumours > 5 cm maximum dimension.

2. Labour intensive and therefore relatively cost inefficient.

3. Requires a general anaesthetic procedure and inpatient care.

4. Requires operator expertise.

9.5 **Implant procedure**

9.5.1 **Sources**

Breast brachytherapy can be undertaken with a low dose-rate technique or high dose-rate afterloading.

(a) Low dose-rate uses iridium wire cut to specific lengths and afterloaded into either flexible plastic tubes or rigid steel needles.

(b) High dose-rate afterloading uses a standard commercial afterloading machine with an iridium source. A bridge-type template with rigid needles is usually employed, available from the standard manufacturers. An example is shown in Fig. 9.1.

For most settings the high dose-rate (HDR) afterloading technique will be employed reducing staff exposure and the need for prolonged patient isolation. There may be instances where it is considered that a low dose-rate (LDR) implant has biological advantages, for example in the re-treatment of a previously irradiated site. Judicious choice of HDR fractionation, however, may easily overcome this.

9.5.2 **Volume definition**

Volume definition presents a similar problem to that of a breast boost with external beam treatment. This should be identified pre-implant by a combination of clinical examination, reference to preceding clinical description, operative description, if relevant, and mammography. Clips placed at surgical excision may be of value, but there is evidence that these can migrate, particularly if several weeks have elapsed during which external beam treatment has been given. Ultrasound and computed tomography (CT) may give further information.

Prior to the procedure, the implant volume (area) corresponding to the CTV should be marked on the patient's skin with indelible marker pen to ensure that the correct area is implanted when the patient is under the effect of the anaesthetic.

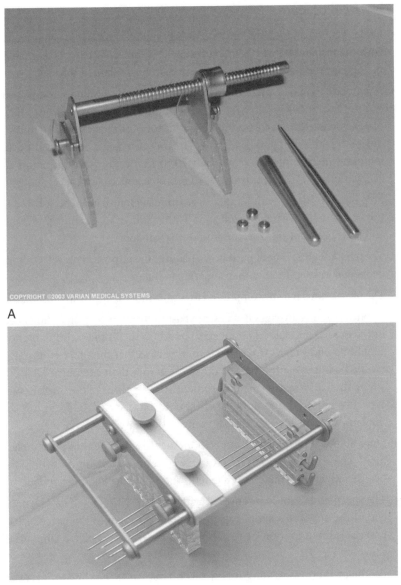

A

B

Fig. 9.1 Template and bridge for breast implant (A) and an alternative design with rigid needle applicators *in situ* (B).

Operative procedure

- Under general anaesthesia the patient is best placed supine or semi-supine with the ipsilateral arm abducted and held on an arm board. This enables the operator to have free access to the area to be implanted.

- A template should be used whether an LDR or HDR technique is employed. It is not acceptable in current practice to implant the breast without a template to define the source positions. Applicator positions, as defined in Paris dosimetry, either in parallel rows of squares or triangles is defined by the template and the spacing is typically 10 to 15 mm; most of the commercially available HDR templates use 10 mm spacing.

- Free hand implants with templates require the template to be sutured to the skin. A bridge arrangement such as that shown for HDR is preferable and far more stable for the duration of the implant.

- The template should be aligned in the plane of the implant. This will typically be medial to lateral rather than superior to inferior.

- The breast tissue may be compressed to form a more stable treatment volume.

- Applicators, usually rigid needles, are passed through corresponding holes in the medial and lateral template to build up the implant. Implantation should be undertaken from lateral to medial so that the sharp ends of the needle lie in the central part of the chest, rather than extending laterally, where they may restrict arm movement. Two or three plane implants are typically required and applicators should be inserted building up the implant from the lower or deeper plane to the more superficial planes.

- Once all the applicators are in position, they are fixed within the template by a locking mechanism. It is best to have the applicator central within the bridge with approximately even lengths extending beyond the template.

- The sharp ends should be covered with a plastic cap for safety. The open end of the applicator should also be capped throughout the procedure to prevent contamination of the source channel. A completed implant is shown in Fig. 9.2.

- Padding with dressings and support with tape across the chest wall is often useful, taking care however, that previously irradiated areas of skin are not used for taping, since, if fragile, this may cause further desquamation.

- An alternative means of breast implantation has recently been introduced and is under evaluation. This uses a single line source within an HDR catheter which is placed in the tumour cavity at the time of local excision. It is retained by an inflatable balloon which acts as a tissue expander within

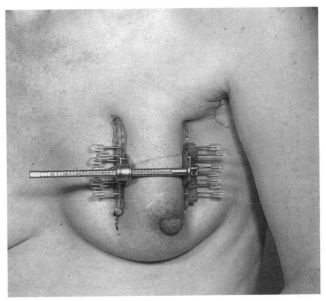

Fig. 9.2 Breast implant *in situ* (see in colour, Plate 12).

the cavity. Once the wound has settled, typically 7 to 10 days after surgery, treatment is delivered through the HDR catheter.

Implant reconstruction

At the end of the procedure in the theatre, it is important to define the applicator positions within the templates. This requires a series of measurements as shown in Fig. 9.3. These should be measured accurately with a steel ruler so that the volume can be defined between the plates of the bridge template and that the wire length or dwell positions to cover this volume can be accurately located. It also enables quality assurance checks to be made to ensure that there has been no movement of the applicators within the templates.

- Radiographic reconstruction where a rigid bridge is used has limited value. Often the bridge structure itself obscures applicators and there are no additional measurements made from the radiographs, that cannot be made on the patient.

- CT dosimetry has been described for breast implant; it has the advantage of more accurate soft tissue definition and the opportunity to customise dosimetry using HDR afterloading.

- If flexible tubing has been used for afterloading of iridium wire then orthogonal X-rays should be taken using dummy wire to identify the positions of

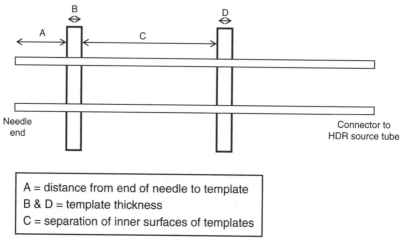

A = distance from end of needle to template
B & D = template thickness
C = separation of inner surfaces of templates

Fig. 9.3 Measurements to be taken after implant.

the catheters, or alternatively, CT scanning provides a more accurate assessment of the catheter positions within the breast and surrounding normal tissue.

9.6 **Dosimetry**

- The use of a rigid template and rigid applicator means that perfect implant geometry is maintained for these implants. Where flexible catheters have been used for LDR iridium wire afterloading, some variation across the length of the plastic tubing may occur and needs to be taken account of.

- The volume is defined between the inner surface of the plates of the template. Typically, unless there is skin infiltration with tumour, 5 mm of skin sparing can be designed into the implant dosimetry.

- Having reconstructed the implant volume, transverse and longitudinal outlines can be defined. With HDR afterloading commercial software is used to define dwell positions along the length of the implant. The source origin is defined from the measurements taken of the applicators within the template, relating the distal end to the template surfaces as shown in Fig. 9.3.

- If LDR iridium wire is to be used, then dosimetry is calculated according to Paris rules since this is a site where crossed ends cannot be achieved, and Manchester dosimetry is inappropriate. Using an HDR afterloading dosimetry programme however, rigid application of Paris or Manchester rules is not essential and an optimized distribution may be obtained. This often results in weighting of dwell times at the extreme ends of the implant to achieve uniformity, adopting the principles of Manchester dosimetry. An example is shown in Fig. 9.4.

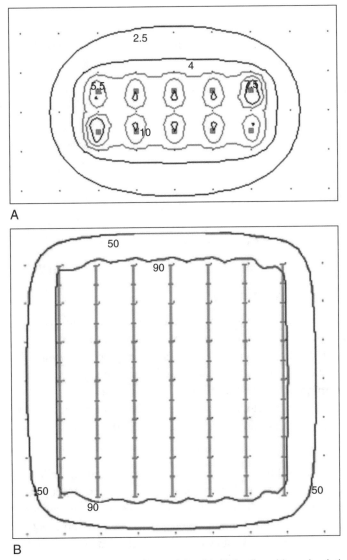

Fig. 9.4 Breast implant dosimetry using weighted end dwell positions simulating Manchester dosimetry system.

9.7 **Dose prescription**

(a) As a boost after external beam radiotherapy

Following an external beam dose to the whole breast of 50 Gy in 25 fractions or its equivalent the following dose prescriptions are used.

LDR 25 to 30 Gy at 0.5 Gy per hour

HDR 10 Gy single dose

16.5 Gy in 3 daily fractions (5.5 Gy per fraction)

18 to 22 Gy in 9 to 10 fractions over 3 to 5 days (2 to 2.5 Gy per fraction)

(b) As a primary treatment in the palliative setting

Where locally advanced or recurrent disease is being treated by brachytherapy, the dose prescription depends upon a number of factors.

- Implantation alone for locally advanced disease encompassable within the implant.

LDR 60 Gy at 0.5 Gy per hour

HDR 30 to 36 Gy in 5 to 6 fractions

- Locally recurrent disease having previously received 50 to 60 Gy

LDR 30 to 40 Gy at 0.5 Gy per hour

HDR 10 Gy single fraction or 18 to 24 Gy in 3 to 4 fractions

9.8 **Treatment delivery**

HDR treatments will be delivered in the afterloading room. The applicators will be connected by source delivery tubes. The spacing of the templates should be measured and the distance of the applicators or needles beyond the medial template should be measured also to confirm that the implant has not moved since insertion or the previous fraction.

The dwell times would have been transferred to the control software of the afterloading machine and treatment delivered in the usual way.

LDR

- The patient will be in a protected room with appropriate radio-protection precautions in place including access to self-contained toilet arrangements, radiation protective barriers at the door and a sign defining the time to be spent with the patient in any one day. Visiting restrictions will be clearly defined. The pre-cut iridium wire will be afterloaded into each of the applicators using long-handled forceps. Checks should be made on the implant at least twice daily to ensure that the applicators have not moved using either marks against the afterloading tube or against the skin to verify their position.

- Loading should be timed with a view to implant removal in order that the implant is removed within normal working hours.
- At removal, the live iridium wire should be held with long-handled forceps and placed in a lead-lined box to be transferred back to the sealed source room for disposal. The templates of the implant can then be removed and the applicators withdrawn.

9.9 **Patient care**

Patients having an HDR implant in position may require hospitalization for two or three days, if more than one fraction is to be given. The implant should be supported but the patients need not be confined to bed and can be normally ambulant about the ward. They may require simple to moderate analgesia.

Similarly, an LDR implant should be supported and simple to moderate strength analgesia may be required. The patient will however, be confined to the radio-protected room throughout.

9.10 **Implant removal**

As mentioned above, LDR implant removal should be timed to occur within the working day. For both types of implants, removal is simple. Further anaesthesia is not required. The rigid needles or applicators should be withdrawn in turn following which the templates will be freed, releasing the underlying breast tissue. There is often extensive oedema and indentation around the site of the templates as shown in Fig. 9.5. The patient should be reassured that within a few days this will resolve and the breast will resume its normal shape and appearance other than the small puncture marks at the site of applicator insertion. No additional treatment for this is required.

9.11 **Clinical results**

No direct comparison of boost by brachytherapy compared to boost by external beam treatment for the management of breast cancer has been undertaken. There are a number of individual series all of which attest to the efficacy of brachytherapy with comparable results to those using external beam treatment. The evidence for breast boost radiotherapy is best defined by the EORTC breast boost trial in which 5569 patients having breast conserving surgery were randomized to receive a breast boost following 50 Gy in 25 fractions external beam whole breast treatment. About 8% of the boost group had an interstitial implant boost using iridium at a dose-rate of 10 Gy per day to a dose of 16 Gy. There was an absolute reduction of 2% in local recurrence with a boost representing a hazard ratio of 0.59; the effect was greatest in women under 40 years where the absolute reduction was 9.3% with a hazard ratio of 0.46.

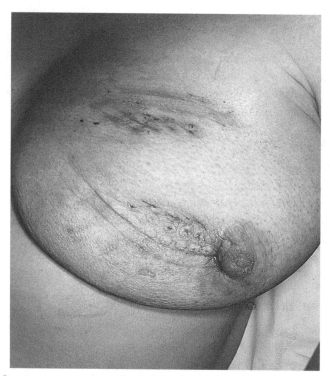

Fig. 9.5 Breast appearance immediately following implant removal with oedema and indentation from bridge sites against skin (see in colour, Plate 13).

Similarly there is no randomized comparison between HDR and LDR brachytherapy for breast boost implantation. One case control study has been published of 575 patients, who after an external beam dose of 50.4 Gy in 28 daily fractions either received an LDR boost of 20 to 30 Gy or an HDR boost of 18 to 22 Gy in 9 to 10 fractions. Patients were matched for age, size, nodes, intraduct element, receptor status, and the use of adjuvant hormones or chemotherapy. No significant difference between the two types of boost for local control and level of cosmesis was seen, 82% of the LDR group and 95% of the HDR group reported excellent or good cosmesis and 87.5% of the LDR group and 91.4% of the HDR group had sustained local control.

Further reading

Van Limbergen E, Mazeron JJ (2002). Breast cancer. In: Gerbaulet A, Potter R, Mazeron JJ, Meertens H, Evan Limbergen (eds.). *The GEC ESTRO handbook of brachytherapy*, pp 435–54. Brussels: ESTRO.

Bartelink H, Horiot JC, Poortmans P *et al.* (2001). Recurrence rates after treatment of breast cancer with standard radiotherapy with or without additional radiation. *NEJM* **345**:1378–1387.

White JR, Wilson JF (2001). Low dose rate brachytherapy for breast cancer. In: Joslin CA, Flynn A, Hall E (ed.). *Principles and practice of brachytherapy using afterloading systems,* pp 343–53. London: Arnold.

Vicini F, Jaffray D, Horwitz E *et al.* (1998). Implementation of 3D-virtual brachytherapy in the management of breast cancer: a description of a new method of interstitial brachytherapy. *International Journal of Radiation Oncology, Biology and Physics* **40**:629–635.

Mazeron JJ, Simon JM, Crook J *et al.* (1991). Influence of dose rate on local control of breast carcinoma treated by external beam irradiation plus iridium 192 implant. *International Journal of Radiation Oncology, Biology and Physics* **21**:1173–77.

Chapter 10

The role of brachytheraphy in miscellaneous sites

Catherine Coyle

This chapter overviews the role of brachytherapy in less common sites: brain, bile duct, penis, skin, and sarcoma.

10.1 Brain

The management of brain tumours is complex and emotionally draining, requiring full multidisciplinary team commitment. Although they only account for less than 2% of all malignancies, the impact of primary central nervous system (CNS) tumours on the patient and carers is profound. Around 20% of all paediatric malignancies are CNS, but the role of brachytherapy is better established in adults. Historic use of interstitial implants for the management of brain tumours can be traced back to the 1920s.

10.1.1 Indications

1. Primary Malignant Cerebral Tumours—de novo
2. Primary Malignant Cerebral tumours—recurrent
3. Meningiomas
4. Cerebral Metastases

Primary malignant cerebral tumours have several characteristics that are radiobiologically important.

1. They rarely metastasize.
2. Imaging techniques with MR allow better localization.
3. There is little target movement within the skull.
4. They require a high dose of radiotherapy.
5. Recurrences are mostly within 2 cm of the target.
6. There is a high hypoxic cell ratio.
7. There is close proximity of important organs at risk.

10.1.2 **Technique**

Although previously orthogonal films were utilized, now CT and particularly MR scans are used to define the target. A stereotactic frame is fixed to the skull and, after volume acquisition and target delineation, a three dimensional attempt at dosimetric calculation is made. This includes the optimal position, trajectory, and number of catheters—the fewer the better. The virtual pre-plan is then carried out under the same stereotactic conditions and further scanning performed to assess the position before the catheters are loaded. Iridium 192 is the commonest isotope used in Europe. The dose-rate requested is 45–55 cGy per hour and the dose prescribed depends on the previous external beam radiotherapy dose and the proximity to organs at risk. Some centres have experience of per operative iodine seed placement and some with high dose-rate (HDR) brachytherapy, but the re-operation rate for symptomatic necrosis was very high.

The patients require steroids and routine anti-convulsants, with a significant risk of brain necrosis as the most common side effect. The radiological follow-up of patients can be confusing because of necrosis, and newer modalities such as PET scanning or MR spectroscopy are being explored. Infection or bleeding due to catheter placement is low.

While survival is clearly an important endpoint, in this group of patients, quality of life is at least as valuable. Although several series have claimed increased survival in newly diagnosed glioblastoma with the addition of brachytherapy, the median survival remains dismal.

10.2 **Sarcoma**

Soft tissue sarcomas account for less than 1% of all adult neoplasms. Although they can affect soft tissues anywhere in the body, the most frequent sites are the extremities. They are managed within central multi-disciplinary teams, potentially by any combination of chemotherapy, appropriate functional or limb preserving surgery, and radiotherapy.

Theoretical advantages of brachytherapy are:

1. Less extensive surgery.
2. Synchronous brachytherapy will allow radiation in oxygenated tissues prior to hypoxic scarring.
3. Direct visualization of the tumour bed reducing geographical miss.
4. Short treatment time.
5. Feasibility, even when surgery and/or external beam radiotherapy have previously failed.

The surgical technique consists of *en bloc* resection of tumour with wide margins, including all previous incisions and biopsy paths. The target includes the tumour and 2–5 cm margins in each direction, varying the margins depending on anatomical boundaries. A single plane implant is usually sufficient and the number of afterloaded tubes are determined by Paris rules. The points of needle insertions are marked on the skin and the tubes are inserted prior to reconstruction and wound closure.

10.2.1 Technique

Rigid needles are spaced uniformly and embedded in the depth of the post-operative field. Stiff nylon cord is passed through the needle, which is then removed. Hollow tubing is then threaded over the nylon cord, clipped at the very end and then pulled through the wound. The cord is then removed and the hollow tube ends snipped to remove any damaged portion. Crimped buttons are used to stabilize the implant until the surgical procedure is completed. Particular attention is needed towards wound closure. To reduce the risk of wound complications, the tubes are not afterloaded for several days post-operatively.

Image guided dosimetry is attractive in this scenario and efforts with both CT and MR dosimetry calculations are under research. The dose depends on the clinical situation and the state of the histopathological margins (Fig. 10.1).

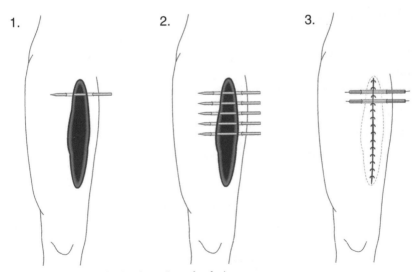

Fig. 10.1 Technique for implantation of soft tissue sarcomas.

In paediatric oncology, brachytherapy shows added advantages by giving a high dose of radiotherapy locally and reducing impaired bone and soft tissue growth.

The Gustave Roussey experience mostly treated rhabdomyosarcomas in head and neck and pelvis:

Primary treatment: chemosensitive tumours received chemotherapy and the radiation component was achieved by brachytherapy, rather than external beam, in the vast majority of patients.

Secondary treatment: a number of patients in the IGR series received brachytherapy on isolated relapse.

10.2.2 Results

Local relapse rates of around 80% are reported.

10.3 Penis

With the advent of advanced reconstructive surgical techniques, penile brachytherapy has become a less common treatment for this cancer that accounts for only 1% of male cancers. Ninety five percent are squamous cancers and, typically, they occur in uncircumcised males.

In the treatment of penile cancer with an implant the patient should first be circumcised so as:

1. to allow for acute tumour extension;
2. to reduce tumour volume;
3. to reduce morbidity;
4. to allow follow up.

Metastatic disease to the inguinal nodes is relatively common, and inguinal lymphadenopathy requires cytological sampling to distinguish infection from tumour.

Indications for brachytherapy include:

1. early stage disease;
2. tumour size less than 4 cm;
3. less than 1 cm invasion of the corpora cavernosum.

10.3.1 Technique

Under anaesthesia the patient is catheterized. The implant is usually two plane with 1–1.5 cm separations. Usually, around six lines are required to make a type of scaffolding as shown in Fig. 10.2. The dose is 60 Gy to the reference iso dose. Foam protection and lead shielding for the gonads and skin is required as shown in Fig. 10.3.

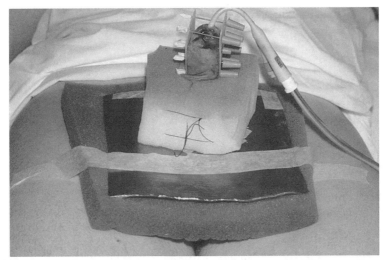

Fig. 10.2 Penile implant scaffold effect (see in colour, Plate 14).

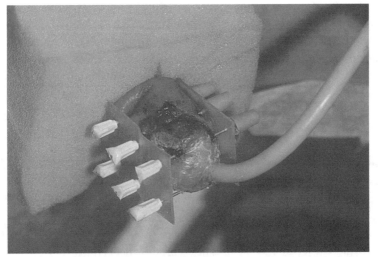

Fig. 10.3 Penile implant with foam and lead protection of the testicles (see in colour, Plate 15).

10.3.2 **Complications**

1. Urethral stricture

2. Ulceration

3. Necrosis

4. Pain

5. Oedema

6. Impotence

10.3.3 **Results**

In early tumours, a disease specific survival at 5 years of 85% is to be expected.

10.4 **Skin**

With the development of modern plastic surgery techniques, brachytherapy has a limited role to play in the management of skin cancer. Occasionally a mould technique is considered in elderly patients with squamous cancers on the dorsum of the hand or the lower limb. A perspex cast is made and the high dose-rate catheter sources are distributed over the PTV following the Patterson Parker rules. Typical fractionation is 45 Gy in 10 fractions, although, in the lower limb, more prolonged fractionation may be advisable.

Keloid scars are more typical in darker skin populations. They may result from therapeutic scarring, i.e. surgical, cosmetic trauma, i.e. earrings, or accidental, i.e. gun shot wounds. As they can be visible and unsightly, plastic surgeons are often consulted about removal. This can be accompanied by the use of silicon gel, pressure therapy on earlobes or intralesional steroids.

However, the exuberant scar formation has a tendency to recur following excision and there is a role for radiation to reduce this tendency. Because the aim is to inhibit fibroblast activity, radiotherapy is given as soon as possible post-resection of scar. In the case of brachytherapy, a fine plastic tubing is implanted into the scar bed preoperatively. The surgeon needs to ensure that the tube itself is not sewn into the scar. The brachytherapist then inserts the appropriate length of iridium wire at their earliest convenience. Unlike other brachytherapy techniques, the iridium is brought deliberately to the surface of the lateral edges of the scar and is visible within the tube to ensure coverage along the whole scar. An alternative technique uses an HDR afterloading catheter or flexible interstitial applicator passed along the scar bed.

This is a particularly useful technique for difficult surfaces, such as behind the ear or around the neck. The dose is usually around 15 Gy at 5 mm. The patients must still be consented on the background of a small risk of tumour induction at a much later date, and careful consideration must be given to sites near the thyroid in a young age group.

10.5 **Bile duct**

These rare tumours present late and have a poor prognosis. Occasionally, cure is achieved by radical resection but the vast majority of patients are considered unsuitable for surgery. Non-surgical treatments, including chemotherapy and radiotherapy, are the focus of increasing interest. Many patients present with obstructive jaundice and their biliary lesion may be traversable by stenting. The brachytherapy procedure is performed with an experienced endoscopist or interventional radiologist.

10.5.1 **Indications**

1. As the primary palliative procedure
2. In combination with external beam radiotherapy.

10.5.2 **Technique**

There are essentially two approaches:

1. Endoscopic
2. Trans-hepatic.

The endoscopic technique relies on initial endoscopic retrograde cholangiopancreatography. The stricture is then identified, a guidewire advanced through and a Seldinger technique used to get a tube in position. The source train is then advanced to the correct position, all under fluoroscopic control, and then stricture plus margin are covered by the appropriate dwell positions in the remote HDR afterloaded source train. Potential problems include inability to get sufficiently long source train and inability to manoeuvre the source train around tight angles. This makes for a difficult procedure, although, it may be improved by further miniaturization of sources.

The trans-hepatic route requires an interested interventionist. Reliance on cholangiography to outline the CTV and perform measurements fluorscopically, has been replaced by the interventionist placing the trans-hepatic stent and giving detailed calculations to the external surface. The brachytherapist uses a hollow tube to place the sources at the appropriate distance from the external surface.

Dose varies according to the use of external beam radiotherapy and intent. Typical doses are 5 Gy per fraction, prescribed at 10 mm from the catheter, with 4 fractions over 48 hours.

10.5.3 **Complications**

1. Infection: antibiotic prophylaxis

2. Haemorrhage

3. Blockage of stent or drain

10.5.4 **Results**

Median survivals of 10 months have been reported.

Further reading

Nag S (ed.). (1994). *High dose rate brachytherapy: a textbook*. Futura publishing.

Harrison LB, Franzèse F, Gaynor JJ, Brennan MF (1993). Long terms results of a prospective randomized trial of adjuvant brachytherapy in the management of completely resected soft tissues sarcomas of the extremity and superficial trunk. *International Journal of Radiation Oncology, Biology and Physics* **27**:259–65.

Hilaris BS, Bodner WR, Mastoras CA. Role of brachytherapy in adult soft tissue sarcomas. *Sem Surg Oncol* **13**:196–203.

Gerbaulet A. (2002).Tumours of the penis. In: Souhmi RL, Tannock I, Hohenberger P, Horiot JC (ed.). *Oxford textbook of oncology*, 2nd edn, pp 2047–56.Oxford: Oxford University Press.

Kiltie AE, Clwell C, Close HJ, Ash DV (2000). Iridium 192 implantation for node negative carcinoma of the penis; the Cookridge Hospital experience. *Clinical Oncology* **12**:25–31.

Neave F, Neal AJ, Hoskin PJ, Hope-Stone H (1993). Carcinoma of the penis: a retrospective review of treatment with iridium mold and external beam irradiation. *Clinical Oncology* **3**:207–10.

Escaramant P, Zimmermann S, *et al.* (1993). The treatment of 783 keloid scars by iridium 192 interstitial irradiation after surgical excision. *International Journal of Radiation Oncology, Biology and Physics* **26**:245–51.

Chapter 11

Endovascular brachytherapy

Catherine Coyle

Coronary artery and peripheral vascular disease are major world health concerns. Sophisticated invasive techniques such as vessel wall stenting are increasingly utilized. Radiation treatment to impair neo-intimal proliferation is a rapidly expanding area of interest. Angioplasty and stenting, performed by interventional radiologists, has an in stent re-stenosis rate of 30 to 50%. It has been shown that under the guidance of angiography, and in some cases intravascular ultrasound (IVUS), low doses of gamma or beta irradiation can be used to significantly reduce the stenosis rate. It has been pointed out that intravascular irradiation is not a compensation for a technically poor angioplasty, and the angioplasty must be optimized before the brachytherapy is carried out. There are already multiple, randomized single, and multi-institutional studies evaluating many aspects of vascular brachytherapy. This includes alternatives such as radioactive coated stents and drug eluting stents. The latter have become increasingly popular and in Europe, have superceded endovascular brachytherapy significantly. There is also an interest in external beam radiotherapy, particularly in arteriovenous haemodialysis shunt management.

The causes of restenosis are complex but include remodelling in the artery wall and neo-intimal hyperplasia as a response to vessel wall injury. Stenting itself may increase the lumen mechanically, but by causing damage, may stimulate neo-intimal hyperplasia, platelet aggregation, thrombus formation, activation of macrophages, and smooth muscle cell proliferation. A number of studies indicate that radiation treatment reduces vascular lesion formation after ballooon over-stretch injury.

Relative contraindications to brachytherapy are, de novo lesions in diabetic patients, multiple vessel angioplasty, lesions at bifurcation of vessels, de novo small vessel disease, previous chest wall radiotherapy and previous, significant endovascular brachytherapy to the same segment.

In 2001, a European group drew up recommendations for prescribing, recording, and reporting in endovascular brachytherapy. The target definition for brachytherapy is determined by pre- and post-procedure angioplasty.

Therefore, the lesion length and the interventional length are decided before-hand at angiography by the interventional radiologist. Adequate margins must be used to prevent the 'edge effect' of restenosis at the edge of the stent.

The American Association of Physicists in Medicine has issued a task group report TG 60 on Intravascular Brachytherapy Physics, where for catheter based systems, the dose at 2mm from the source is specified and for radioactive stents the dose should be specified at a distance of 0.5 mm from the surface. It demands that the TG 43 dose calculation specification of air kerma strength is given for gamma sources and beta sources and should be expressed in terms of dose-rate in water. It specifies a robust quality assurance programme and the components of an appropriate multidisciplinary team.

11.1 **Radiotherapy definitions**

Clinical target volume: volume, length, and depth of the angioplasty balloon injured part of the vessel wall. This must be longer than the interventional length.

Planning target volume: must cover the CTV plus positional margin to allow for inaccuracies of movement of source, particularly with adjacent cardiac motion.

Reference depth: with brachytherapy it is known that prescription should be to a specified depth and therefore, having a reference depth would be useful. This is 1 mm in the coronary and 2 mm in the peripheral vessels.

Treated volume: this should be as close to the PTV as possible but may vary with commercial devices.

Irradiated length/depth/volume: While endovascular brachytherapy is a relatively new indication, records should be kept for longer term studies.

See Fig. 11.1 for definitions for endovascular brachytherapy.

11.2 **Sources**

A number of commercial devices are now available, utilizing either beta or gamma irradiation. A new range of drug or irradiation coated stents are also being evaluated. For peripheral arteries, it is thought that gamma irradiation may be more useful with increasing depth as the diameter is around 6 mm.

Ideally:

1. The source needs to be 0.5 mm in diameter, but stiff enough to be able to be introduced via a femoral approach nearly one metre away and yet flexible enough to negotiate bends in the coronary arteries.

2. The source needs to be able to treat a target length of 3–4 cm of arterial wall of 2–6 mm diameter and with wall thickness of less than 3 mm.

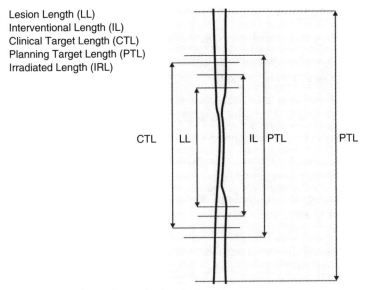

Lesion Length (LL)
Interventional Length (IL)
Clinical Target Length (CTL)
Planning Target Length (PTL)
Irradiated Length (IRL)

CTL   LL   IL | PTL   PTL

Fig. 11.1 Definitions for endovascular brachytheraphy.

3. There must be very steep dose gradient to minimize dose to the normal vessels and myocardium.

4. The treatment must be given in a single fraction with a sufficiently high dose-rate to keep the treatment time below 15 min in order to reduce trauma and subsequent thromboembolism.

It should be noted that unlike oncological brachytherapy, the target is millimetres rather than centimetres away from the source. There are controversies over the need for centring the source within the eccentric blood vessel lumen, with data supporting both a policy of ignoring this, and of self-centring catheters.

The prescription dose depends on the prescribing point but is generally around 14 to 18 Gy. The prescription point must be clearly stated to compare the literature. Restenosis at the stent edge is the major drawback, requiring careful calculation of the safety margin. Appropriate safety margins will also reduce geographic miss. It is important that the entire balloon damaged part of the vessel is adequately covered.

The long term safety of endovascular brachytherapy is uncertain at this point. Questions about tumour induction and aneurysm formation remain unanswered. There is an increased risk of late thromboembolism relating to vessel wall injury, and to reduce this all patients are prescribed 6 months of Clopidogrel. Late thrombotic events at about six months have been noted particularly, in association with the combination of new stenting and brachytherapy.

11.3 **Results**

The most significant trials include:

- ◆ SCRIPPS where the three year restenosis rate dropped from 48 to 15%, using a non-centred iridium 192 (Ir 192) device.
- ◆ WRIST also used Ir 192 non-centred intracoronary brachytherapy with six months stenosis drop from 58 to 19%.
- ◆ Beta WRIST looked at centred Y90.
- ◆ INHIBIT was a multicentred trial using Phosphorous 32 (P32) and at nine months the rate came down to 26% from 52%.

The trials for femoropopliteal use include five Vienna trials and the Paris trial. All used iridium 192 HDR.

Endovascular brachytherapy still has to gain recognition outside of major US centres and the very extensive Vienna experience. The increasing use of brachytherapy in treatment of benign disease will require careful long term follow up.

Further reading

Potter R, Van Limbergen E, Dries W *et al.* (2001). Prescribing, recording and reporting in endovascular brachytherapy. Quality assurance, equipment, personnel and education. *Radiother Oncol* **59**(3):339–60.

Waksman Ron (ed.). *Vascular brachytherapy*, 2nd edn. Futura Publishing, USA, 2002 (ISBN 0879934891).

Chapter 12

Quality assurance

Peter Bownes

12.1 Introduction

The main objective of any brachytherapy treatment is optimization of the dose distribution to the patient, so that the prescribed dose is delivered to the prescribed site within acceptable constraints, while limiting the dose to normal tissues outside the target volume. This allows the treatment to be delivered in accordance with the ALARP (As Low As Reasonably Practicable) principle.

A quality assurance (QA) programme is required to ensure that these objectives are met with and that the equipment continues to exhibit satisfactory performance, as defined, when the results from the quality assurance procedure fall within an acceptable tolerance. This tolerance should guarantee a high level of safety regarding the patient and staff and an acceptable accuracy of dose delivery, in terms of positional and temporal accuracy.

The medical physics expert (MPE) should be responsible for the management of the QA programme and for the acceptance testing of new or modified equipment, its commissioning, and definitive calibration. The goal of the QA programme is to maximize the likelihood that all treatments are delivered correctly and that the IR(ME)R practitioner's clinical intent is executed safely, with regard to the patient and others. This should address every step of the treatment delivery process:

◆ Diagnosis and treatment decisions

◆ Implant design and applicator/source insertion process

◆ Definition of target volumes and normal tissue structures

◆ Reconstruction of applicator positions

◆ Treatment planning process

◆ Treatment delivery process

The quality assurance programme must ensure that all equipment operates safely and correctly, parameters used in dosimetry calculations are correct, (e.g. radioactive source strength and geometric source reconstruction), and all sources are accounted for. It is also important that test and measurement equipment should have its own quality control to ensure the accuracy of subsequent QA tests. Documentation of all QA should be kept for future reference, in order to:

- Perform trend analysis to pick out patterns, and predict problems before they occur and apply preventative action.

- Provide evidence of previous quality assurance sessions and highlight any previous problems.

- Keep track of servicing, including parts that have been replaced or upgraded.

All quality assurance procedure should consider this general format:

- What is the **objective** of the quality assurance procedure?

- What is the **frequency** of the procedure, and when and how should it be performed?

- A full **description** of the procedure in terms of detailed instructions should be included, along with actions to take if the test is outside of the tolerance limits.

- All **results**, with values and ranges, which should be **documented** and signed.

IPEM Report 81[1] provides UK with recommendations for the quality control required in brachytherapy with additional recommendations given in the BIR/IPSM recommendations for brachytherapy[2] and IAEA TECDOC-1274[3]. Quality assurance issues for afterloading units are also covered extensively in other textbooks[4–6] and in the AAPM task group reports 41, 56 and 53[7–9]. The remit of this chapter is to give a general overview of relevant QA procedures to brachytherapy.

12.2 Manual afterloading techniques

Many brachytherapy procedures use manual afterloading techniques. Empty applicators or catheters are inserted into the tissues or in a body cavity of the patient and the sources are loaded later. This has the advantage that with no radiation exposure present time can be spent on the geometric position of the applicator. Imaging checks on source positions can be done using dummy sources or simply using the applicator itself to provide contrast on the images. The implant is reconstructed and dosimetry calculations performed. The live radioactive sources are then manually loaded into the localizing applicators, either in their own ward room, or at the end of the procedure, in theatres. Examples of such treatment techniques include iridum-192 wire or hairpin treatments and certain gynaecological applicator systems using caesium-137.

The core QA tests required for manual afterloading brachytherapy sources include:

♦ Verification of source documentation from the manufacturers

♦ Measurement of source strength and reconciliation with manufacturer's documentation

♦ Verification of activity distribution within the source

♦ Leakage testing of the radioactive source

♦ Applicator integrity

♦ Verification of plan applicability

12.2.1 **Source documentation**

Every new source delivered from a manufacturer has a delivery note and source certificate. A standard source certificate should include the following information:

♦ Source serial number

♦ Regulatory requirements manufactured to

♦ Isotope used and its chemical and physical form

♦ Active dimensions of the source

♦ Encapsulation used and the capsule dimensions

♦ Air kerma rate at 1 m and the calibration date

♦ List of tests performed on the source and the results, e.g. leakage test, surface contamination, construction integrity and construction strength.

The entire manufacturer's documentation should be checked on delivery and appended to all local documentation, and the sources should be logged into the local source inventory, and stored in a secure and shielded place. The control of radioactive sources and the need for regular safe audits to check accountability is discussed in Chapter 3.

12.2.2 **Source strength confirmation**

The source strength of all radioactive sources should be independently measured by the user before clinical use. User calibration is essential to verify the vendor's stated calibration. Any measurement system used should be calibrated for the particular source type being measured, and the calibration should be traceable back to a national standard laboratory, for example

National Physics Laboratory (NPL) in the UK or the USA's National Institute of Standards and Technology (NIST).

Direct traceability is established when a source or calibrator has been calibrated at the standards laboratory. Secondary traceability is established when the source is calibrated against a source of the same design, geometry, and comparable strength, which has direct traceability, or when a source has been calibrated using an instrument with direct traceability.

The result of the source strength measurement should be within ±5% of the manufacturers' stated strength of the source. Reference should always be made to the manufacturer's source certificate for dose calculations.

There are three methods of calibrating sources. The most frequently used method, a re-entrant ionization chamber, is mainly used for low dose-rate conventional sources used for manual afterloading; the second method uses a thimble ionization chamber to measure the air kerma rate directly, which is recommended for the measurement of high dose rate sources; the final method uses a thimble ionization chamber in a solid phantom. The calibration of the measuring device is strongly dependent on energy and on the isotope that is being used. Further information on the use of a re-entrant ionization chamber for source strength measurement can be found in Appendix A.

12.2.3 **Autoradiographs**

An autoradiograph is a simple method to check the distribution of radioactivity within a source. The radioactive source is placed directly onto a film for a suitable exposure time to produce an image of the source activity distribution (e.g. a 195 mCi Cs-137 source produces an acceptable image on Kodak Xomat V film in one or two min). When the film is developed, the blackening on the film indicates the distribution of activity within the source, and indicates if they are any unexpected gaps, as in Fig. 12.1, or hot-spots, indicating non-uniform distribution. Autoradiography can also be used to verify the configuration of pre-loaded source chains.

12.2.4 **Leakage testing**

All sealed radioactive sources should be leakage-tested by the manufacturer and the test certificate should be filed for future reference. If no leakage test certificate is provided then the user is required to test for both leakage and surface contamination, before clinical use. Brachytherapy sources are required to be tested for leakage annually or when damage of the source is suspected. Leakage tests consist of wiping the radioactive source using forceps with a swab moistened with water or methanol. The swab is then measured using a sodium

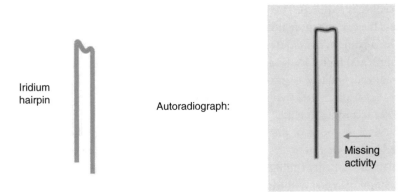

Fig. 12.1 Autoradiograph of iridium-192 hairpin.

iodide scintillation counter. If the activity of the measured swab is less than 200 Bq, then the source is considered leak free.

12.2.5 Other brachytherapy QA checks

Before each clinical use applicators should be visually inspected for any damage. Radiographic checks can also be done for more rigorous checks of applicator integrity. Loading equipment should also be routinely checked to function correctly. For example an iridium-192 wire cutter should be checked to see if there is no significant surface contamination, visually inspected to check correct function; and to check that the wire is cut appropriately and to the correct length. For the iridium wire, the user must also inspect the sealing mechanism of the plastic tubes into which the wire has been placed. Radiation monitors used for radiation safety surveys of areas during and after source use should be calibrated on an annual basis (details of this are out of the scope of this chapter).

12.3 High dose-rate remote afterloading

Advances in afterloading techniques were initially driven by the desire to improve radiation protection aspects of brachytherapy. The use of remote afterloading techniques has extended the radiation protection advantages to the whole of the brachytherapy procedure. Afterloading refers to any method where applicators are placed in the tissues or in the body cavity and the sources are loaded later on. In remote afterloading systems, the radioactive source is contained in a lead safe close to the treatment couch. The source is then transferred into the required position by the machine itself and the source can be temporarily retracted from the patient and stored back in the safe, if any nursing procedure is required. Therefore, no radiation exposure should occur to staff unless the equipment malfunctions or the interlocks fail.

The advances in remote afterloading have allowed sources of increased strength and high specific activity to be used, which allow treatment times to be reduced. Quality assurance will be discussed within this section for high dose-rate iridium-192 stepping source units. The definition of high dose-rate (HDR) according to ICRU [10] is a dose-rate greater than 0.2 Gy min^{-1}. There are several machines of this type available including the microselectron–HDR (Nucletron), Gammamed (Varian) and the Varisource (Varian). The source dimensions vary depending on the manufacturer and model, ranging from 0.2 to 1.3 mm diameter and 1 to 20 mm active length, with typically a nominal air kerma rate of 42 mGy h^{-1} at 1m. The active wire is encapsulated in a stainless steel capsule, laser welded permanently onto a cable drive.

The unit contains a single miniature iridium-192 source that is sequentially stepped through a series of defined dwell times within each applicator. Stepping source brachytherapy offers positional and temporal degrees of freedom, which have allowed the development of conformal brachytherapy. This section covers the quality assurance requirements for the HDR treatment unit and the facility, in general. Good maintenance and quality control are especially vital with HDR systems, because, the dose is delivered at such a high rate (dose-rate in water ~7.5 Gy/min at 1 cm), that there is potential to do a large amount of damage to a patient in just a few minutes, if faults occur.

A typical QA schedule is shown in Table 12.1 and further suggestions are made in IPEM 81[1] and TG56[8].

12.3.1 Source calibration

HDR remote afterloading units use a single iridium-192 source with an air kerma rate up to 42m Gy^{-1} at 1 m (activity up to 370 GBq). The half life of iridium-192 is 73.83 days, and it is normal to replace a source every three months.

The source suppliers provide a calibration certificate with the new source that states the air kerma rate within ±5%, measured with a calibrated system traceable to a national standard. The calibration certificate also includes details of other tests performed on the source, which include:

+ Total length and diameter of the source
+ Visual check on the capsule weld
+ Capsule integrity (applying a force of 15 N)
+ Cable-connector assembly (applying a force of 40 N)
+ Surface contamination—according to ISO9978 wet wipe test
+ Leakage Test—according to ISO9978

Table 12.1 Typical quality assurance schedule for a HDR remote afterloading unit

Frequency	Test	Method
	Machine function tests	Check correct function of all interlocks, indicators, emergency equipment, area radiation monitor, audio, and visual systems.
	Source data Checks	Verify date, time, and source strength, in planning computer and treatment unit.
Pre-Treatment	Positional accuracy	Verify source position of the stepping source either using a source stepping viewer, and CCTV, or with film.
	Temporal accuracy	Check temporal accuracy against a stopwatch.
	Applicator integrity	Check visually for damage. Run through a complete cycle of a simulated treatment.
	Source calibration	Two independent checks on the air-kerma rate of the new source. Data entry into the treatment planning system and treatment unit should be verified.
Quarterly (post service and source change)	Source position set-up and verification	Verify the location of the source in its safe position. Verify source position of the stepping source using both a source stepping viewer with CCTV and a film.
	Pre-treatment tests	As above
Post maintenance		Relevant QA performed depends on what maintenance has been done; consult the brachytherapy medical physics expert for details.
As required	Applicator checks	Any new applicator should be checked using radiographic and autoradiographic techniques to verify integrity of the applicator and source position within it.
Annual	See Table 12.3	
Pre-use and quarterly	Treatment planning system QA	Include global system checks, which incorporates all the steps of the planning process and treatment delivery. See Section 12.5 for further details.

The user must independently verify the calibration of the source before the source can be used clinically. The current method recommended in the UK by the joint BIR/IPSM Working Party[2], is to calibrate the HDR iridium-192 sources in air with a Farmer (0.6 cc) ionisation chamber at a distance of 100 mm from the source. The endpoint of the calibration is to determine the reference air kerma rate of the source within an acceptable tolerance (±5%). The calibration procedure involves two parts, with each part being performed by a different physicist with a different calibrated dosemeter/chamber combination. A detailed discussion of HDR source calibration methods are found in Appendix B.

12.3.2 Machine function tests

Routine checks on the functionality of the treatment unit should be done prior to every treatment session. A series of tests, in the form of a checklist, should be performed to evaluate safety interlocks, warning systems, and emergency equipment. These tests should verify the following:

- Source strength data, date and time within the treatment control unit and planning system
- Communication between the computer control unit and the treatment unit
- Radiation warning signs, lights, and audible time delay are present and function correctly
- All door and safety interlocks
- Emergency return mechanisms, including the emergency treatment report indicating what has been delivered, and what remains to be delivered
- Availability of portable Geiger counter, which is to be checked for proper functioning
- Availability of a shielded storage safe in the event of an emergency
- All applicators and accessories used in that treatment session are tested to ensure integrity of the source transport system, applicator, applicator couplings, and connections. The total length of the applicator plus source transport guide tube is a critical parameter and should be verified by the dummy source, or length gauge to within 1 mm accuracy. Visual inspections should also be done to look for damage or constrictions
- Machine timers with a calibrated stopwatch
- Positional reproducibility of the source; see Section 12.3.3.
- Functionality of the treatment console and printer
- CCTV and patient intercom systems

12.3.3 Positional and temporal tests

Delivering the dose accurately to the patient depends on the source strength at the time of treatment being correct, the timer being precise, and consistent, and correct positioning of the source within the correct applicator.

Temporal accuracy should be checked over the entire treatment time range of the device. A number of dwell positions should be checked to ensure the programmed dwell times are delivered within the appropriate accuracy (±1%). Timer accuracy can be checked with a calibrated stopwatch and the linearity of the system can be checked with the use of an ion chamber.

The transit time of the source and the total transit dose should be checked at the commissioning of the treatment unit to ensure they are insignificant, compared to the expected dose delivered. Transit dose is the dose delivered as the source moves from its safe to the first dwell position, during the transit between dwell positions, and then as it returns to the unit's safe position. The total transit dose is dependent on the transit time, source strength, and number of treatment interruptions. It is useful to include this measurement in the quality assurance schedule to ensure that transit dose remains insignificant. Wong[11] and Sahoo[12] describe methods of measuring the transit dose.

Small positional errors in brachytherapy can result in large errors in dose delivery, due to the high dose gradient around the source. One form of positional error in the high dose-rate system is the accuracy of the source positioning within the catheter. Source positioning should be within ±1 mm of the intended position, and this should be included in the pre-treatment tests. The simplest method for pre-treatment checks is to use a source step viewer and the CCTV system to view the source position. A ruler is attached next to the catheter to verify the position of the source and that it steps correctly between dwell positions (see Fig. 12.2a).

Autoradiographs are a useful means of evaluating the position of multiple source positions, see Fig. 12.2b. The source step viewer technique can be compared to this as a double check at monthly intervals or after source position adjustments. This is also a useful tool to check if the dead space between the first dwell position and the end of the applicator is correct. Radiographs can be used to verify the physical integrity of the applicators and their length. Many applicators also have a set of dummy sources, which are designed to simulate the possible dwell positions during treatment. It is essential that these dummy positions correlate with the real dwell positions, which can be checked by obtaining a radiograph of the dummy sources and comparing it with an autoradiograph of the real dwell positions, (see Fig. 12.3).

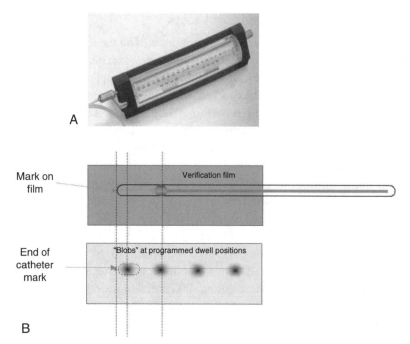

Fig. 12.2 Source position verification. (A) Source step viewer used at Mount Vernon (Reproduced with kind permission of Varian Medical Systems); (B) Autoradiographs of source position.

12.3.4 **Maintenance**

The vendor generally performs planned preventative maintenance on the treatment unit at quarterly source changes, which includes all electrical, electronic, and mechanical component checks. Most vendors also have a rolling replacement programme for parts within the unit, to prevent any errors in function. Many of the major components of the system can be tested with a dummy source. The following components are tested:

- Interlocks and fail-safe/emergency mechanisms
- Control unit/software features
- Radiation monitoring systems both within the unit and the independent monitor
- Power supply and battery backup
- Source drive mechanism, channel indexer and position sensors
- Source transfer tubes and connector couplings
- Source position within the safe
- Length gauge check of the source and cable
- Source position set-up

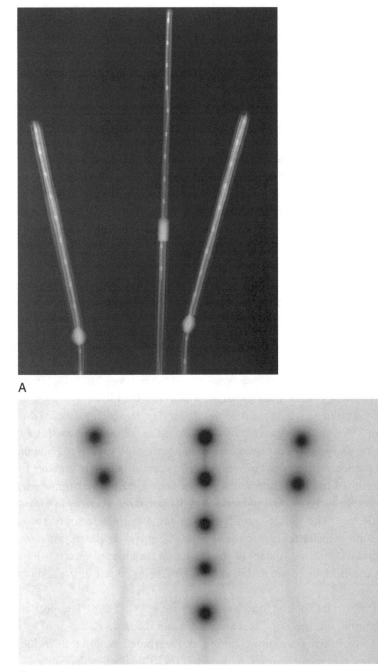

A

B

Fig.12.3 (A) Radiograph of simulated dummy sources within an applicator.
(B) Autoradiograph of 'Real Source' Positions within an applicator.

12.3.5 **Acceptance testing**

Before the acceptance tests are performed, a critical examination needs to be performed on all equipment (installed or erected) or articles that have implications on radiation protection. This examination is required to demonstrate to the purchaser, usually by the installer, that the safety features and warning devices function correctly and that there is sufficient protection from radiation exposure.

Acceptance testing for a new HDR remote afterloading unit must be performed before clinical use. The tests verify whether the device performs to the unit's specifications and regulatory requirements. This allows the user to evaluate the system and provide initial baseline measurements that form the basis of a QA programme. Commissioning tests are performed by the purchaser's representative to ensure the equipment is ready for clinical use, details of commissioning tests are found in Appendix C.

12.4 **Low/medium dose-rate remote afterloading**

The Selectron-LDR (Nucletron) was initially designed for gynaecological intracavitary treatments, but it has also been used for intraluminal and surface mould treatments. The unit contains up to 48 spherical caesium-137 sources of external diameter of 2.5 mm. Source activities between 740 MBq (20 mCi) and 1480 MBq (40 mCi) are available, which with the higher activity sources can give a point A dose in the medium dose-rate range (defined as 2 Gyh^{-1} to 12 Gyh^{-1} [10]. The unit also contains inactive spacers of the same diameter that allow the user to compose source trains of sources and spacers indivdualized to the particular insertion. The sources and spacers are stored in separate compartments of the main safe and when the source trains are composed, the machine places the sources and spacers in the programmed order in the intermediate safe. The source train is then pneumatically driven out through the flexible transfer tubes into the applicator. Sources can be returned to the intermediate safe at any stage of the treatment for planned interruptions or alarm conditions and are withdrawn back to the main safe at the end of the treatment.

The Selectron LDR unit functions in different ways to the HDR remote afterloading unit, but the principles are similar and the main concerns of source calibration, source positioning, timing, and machine function tests including all the relevant safety features, remain the crucial elements of a quality assurance programme.

12.4.1 **Source calibration checks**

The Selectron-LDR has a number of caesium-137 spherical sources with a half-life of 30.17 years, so the same sources will remain in clinical use for a number of years. The sources are required to be calibrated at acceptance and then at least annually thereafter. Calibration is best achieved in a calibrated re-entrant ionization chamber (see Section 12.2.2). Auckett [13] describes an alternative method using a Farmer ionization chamber to measure air kerma rate. Several sources will probably be needed to provide a reasonable dose rate and correction for distance from each source to the chamber will be required. Chamber leakage and background need to be corrected while transit effects have to be avoided.

New sources should be autoradiographed to assess the activity distribution within each source and this should be repeated periodically throughout their clinical use. Films can be visually inspected, using a ruler to compare the film image to the stated active length and to assess the location the activity within the physical capsule.

Sources should be delivered with acceptable leak test certificates. If they have not been tested within the last six months, then they should be tested before use. Leak tests should then be performed annually or if damage is suspected. Once in the unit periodic leak tests can be performed by wiping the inner surfaces of either the source transit tubes or storage containers.

12.4.2 **Machine function tests**

The machine function tests are similar to the HDR system's routine pre-treatment tests (Section 12.3.2) and should therefore include all of the machine's safety features, interlocks, and fail-safe systems. One of the main differences in the systems is the source drive mechanism; the Selectron-LDR uses a pneumatic source transport system, which requires additional checks on the transport tubes and connectors for leaks, constrictions, and other obstacles. Fail-safe mechanisms for air loss and power loss should be regularly tested.

12.4.3 **Positional and temporal checks**

Autoradiographs provide the best method for verification of source position within the applicators and the programmed source train configuration. The accuracy of source position should be achievable to within ±1 mm of the intended [14]. Radiographs of the applicators with a set of dummy sources loaded to simulate the possible source positions should be compared against an autoradiograph of the applicator with the live sources loaded. The applicator tip should be marked to ensure that the dummy sources accurately represent the real source positions to within ±1 mm.

The treatment time mechanisms must be tested for accuracy, linearity, and the transit time of the sources. The timer accuracy should be achievable to within ±1%[14]. Linearity and transit time can be tested with the use of an ionization chamber, taking several measurements for different treatment times [14].

12.4.4 Acceptance testing

Acceptance and commissioning tests are required to ensure the system is operating safely, accurately, and as specified. The tests establish baseline performance parameters, which can be used in routine quality assurance checks to verify the performance of the equipment. They can be divided into:

+ Radiation safety surveys
+ Mechanical and electrical operations of the remote afterloader including the emergency and fail-safe mechanisms.
+ Radiation monitors and other facility equipment should be checked to be operating correctly
+ Source calibration, leak testing, and verification of source activity distribution
+ Integrity of applicators and their connections to the unit
+ Verification of source transport mechanism
+ Temporal and positional checks
+ Verification of programming a treatment
+ Treatment planning system and data transfer verification
+ Documentation and training

12.5 Treatment planning systems

In recent years, the treatment planning process in brachytherapy has increased in complexity as a result of technological advances in imaging, HDR remote afterloading and computer power. Comprehensive commissioning of the system should cover all aspects of the treatment planning process and once in clinical use, regular QA checks are required on the system to identify any data corruption or degradation of the hardware. In addition, if any software upgrades are implemented, or changes to data within the system are made, then these parts of the system need to be recommissioned. Results from the initial commissioning tests provide the baseline records against which routine QA results can be compared. Treatment planning system quality control should be designed to ensure the quality of treatment plans, minimize the possibility of systematic errors introduced at the treatment planning stage, and be designed to complement checks on individual treatment plans. General recommendations on commissioning and quality assurance of treatment planning systems are made by TG53[9], Van Dyk[15], and IPEM 81[1].

12.5.1 Acceptance testing

Acceptance tests are performed on the treatment planning system (TPS) to confirm that the system performs to its specification. The user carries out these tests on the system after installation to check the specifications of the computer hardware, peripherals (including dicom conformance if applicable), software features, and benchmark tests, to look at the accuracy of the system in terms of dose calculation and geometric reconstruction.

12.5.2 Commissioning

Commissioning of the system is performed on the entire treatment planning process. Planning for treating a patient is a series of complex interlinked procedures that include:

+ Diagnosis
+ Patient positioning and immobilization of catheters, sources, and patient
+ Anatomy and volume(s) definition
+ Determination of source arrangement and weightings (in terms of dwell time or activity)
+ Dose calculation
+ Plan evaluation
+ Plan implementation including data transfer

When commissioning the TPS, not only should the hardware and software be tested, but also how it interacts with the treatment planning process and how it is used. Uncertainties and errors in the TPS can result in reduced cure probability or serious complications. The initial part of the commissioning process should examine the hardware/software and review the manuals of the system. The brachytherapy dose calculation algorithm should be studied and understood for all sources to be used. The dose algorithm should be investigated before the commissioning process starts so the tests can be customized to examine the specific system. The following questions should be addressed:

+ What is the dosimetry formalism used by the system?
+ Is the source modelled as a point or a line source?
+ Are the correct factors being employed by the system?

- Is the source accurately modelled in terms of size, shape, and encapsulation?
- How does the algorithm calculate the net affect of attenuation and scattering in tissue and how does it account for inhomogenities?
- Can the system utilize CT data?
- Can the brachytherapy plans be added to external beam plans, and if so, how does it incorporate the different radiobiological models?

Most commercial planning systems adopt the AAPM formalism[16] for dose calculations of low energy emitters and iridium-192, but many still have the option to use the BIR/IPSM[2] formalism, based on dose-rate calculations from a reference air kerma rate specification. Currently, commercial systems do not utilize CT data to take account of tissue inhomogeneities or applicator inhomogeneities in the dose algorithm.

The following dosimetric tests should be performed at the commissioning stage:

- Verification of all source data entry into the system.
- Verification that the algorithm works correctly, comparing the system to either measured, manually calculated, or published data. This should be done for a single source and multiple sources (or multiple dwell positions).
- Verification of dose distributions for a single source, multiple sources, and various test implants.
- Testing of optimization software based on ideal implant geometries.

The planning process contains many non-dosimetric aspects, which must be commissioned and routinely evaluated as part of a QA programme. They include:

- Verification of geometric accuracy of input/output peripherals, for example CT (including registration processes, if available), MRI, ultrasound, film scanner, digitizer, or keyboard
- Accuracy of coordinate reconstruction of source and applicators
- Verification of standard applicator libraries
- Tests on the structure contouring tools
- Verification of plan evaluation tools, including dose volume histograms, 2D and 3D dose display tools and plan comparison
- Accuracy of hardcopy outputs
- Verification of plan transfer to treatment unit

Overall system tests should be performed, which test the complete planning process from start to finish to check on the systematic behaviour of the treatment planning process. These should be repeated for a series of standard plans and scenarios.

Van Dyk[15] recommends that an accuracy in dose calculations of ±3% should be achievable at distances of 0.5 cm or more for any source. The BRAPHYQS group of ESTRO[17] provides a new QA programme for geometric reconstruction using the 'Baltas' Phantom. They define a tolerance level of mean deviation to be ≤ ±1 mm or when the 95% confidence limit (= |mean dev| + 2 standard deviation) is ≤ ±2 mm.

The entire commissioning procedure should be independently checked by a second physicist to ensure that no errors exist or parts of the process not tested. User training is required before the system can become clinical to ensure accurate use. The vendor normally provides some user training to review the system, describe the algorithms and their limitations, provide hands on experience, and describe hardware maintenance and housekeeping procedures. The on-going staff training should include system familiarization, supervised training and review of relevant documentation on the system and departmental procedures.

12.5.3 Routine quality assurance

An integral part of the commissioning process is to establish quality assurance procedures for the treatment planning system. The routine QA programme should include a method for verifying:

- The integrity and security of data files used for dose calculation
- Functionality and accuracy of the peripheral devices for data input
- Integrity of system software and its functionality/features
- Accuracy of output devices and data transfer

The frequency of testing depends to some extent on the brachytherapy workload of the department, but if several calculations a week are made then a typical schedule is shown in Table 12.2. For HDR remote afterloading systems the data for each new source must be entered into the planning system and be independently checked by a second physicist.

Security of the TPS, especially of patient and source data, is an important issue to be addressed before the system becomes clinical. Source data must have a high level of security and the integrity of data files should be checked before use by comparing the source data file against a standard copy of the data file held separately to check for data corruption or unauthorized alterations. Other housekeeping issues include having a system in place to check for viruses and routine backup of patient data. Non-current patient data should be archived to appropriate media at regular intervals to avoid disk space being filled.

Table 12.2 A typical QA schedule for a brachytherapy treatment planning system

Check	Frequency
Calculation of a simple standard plan	Daily
Data file integrity	Daily
Consistency of printed plan documentation and data transfer to treatment unit	Every clinical use
Independent plan check	Every clinical use
Verify geometric accuracy of input and output devices. Calculate doses and compare against a baseline distribution	Monthly
Optimization and plan evaluation functions	At least annually
Overall system test	At least annually
Commissioning tests (see 12.5.2)	On the introduction of new software versions or new sources

12.5.4 Treatment plan checking

Hardware problems or uncommon usage of treatment planning systems could result in inaccuracies not detected in commissioning or regular quality assurance checks. Therefore, it is recommended for all plans that an independent check by a second physicist is performed using a manual calculation, generally based on a conventional system of dosage calculations. The integrity of the data transfer to the treatment unit should be independently checked before the treatment is commenced.

12.5.5 In vivo dosimetry

In vivo dosimetry is the only direct method that compares dose in the patient to that calculated by the TPS. As a final check of the overall treatment process, and when new or unusual treatment procedures are implemented, in vivo dosimetry should be performed on a select number of patients or phantoms. Measurements devices suitable for in vivo dosimetry include TLDs, diodes, and small ionization chambers. GAF chromic film and polymer gels may be used for phantom studies. It is essential to ensure the correct parameters and correction factors are applied to the measuring device used. For example, TLDs should be corrected for variation in sensitivity with energy, if individual sensitivity calibrations are not used.

12.6 **Quality management and independent audit**

Quality management of a brachytherapy service is a series of interlinking processes covering both physical and clinical aspects that ensure a high quality service for all patients and treatment processes. Quality control, quality assurance, and audit are the main components that allow best and safest practice to be achieved in brachytherapy. Quality control is the process that regulates the treatment delivery service as a whole and ensures that it conforms to existing standards. Quality assurance is the planned and systematic actions that confirm to a high degree of confidence that the treatment is delivered as intended. Audits play an important role in the assessment of the delivery and outcome. Independent audits by another centre are useful in determining deficiencies in the overall quality control programme. They provide means of formulating best practice and cost effectiveness.

APPENDIX A

Source strength measurement using a re-entrant ionization chamber

The re-entrant (well-type) ionization chamber is used in isotope calibrators to measure source strength, see Fig. 12.4. It is a cylindrical well with an ion collection volume of gas (usually air) that surrounds the source. In order to avoid measurement errors, it is important to ascertain the characteristics of the chamber before use with each source type. The response of the chamber will be dependent on many factors discussed by Williamson *et al.*[18] and include:

- **Isotope** and **energy** of the source being used
- **Source strength** measured
- Response should be **linear** through its measuring range

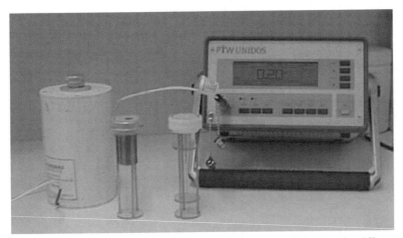

Fig. 12.4 Standard imaging well type (re-entrant) chamber with holders for different sources.

- ◆ **Geometric configuration** of the source and its encapsulation
- ◆ Source **position and orientation** within the chamber
- ◆ **Source holder** used
- ◆ **Effect of scatter** from external shielding

A re-entrant ionization chamber does not provide a direct measurement of source strength, as they require calibration with standard sources traceable to a standards laboratory. The accuracy to which measurements can be performed depends on using calibrated sources with the same geometry and same approximate activity as the sources to be measured. It is important to use holders that allow a reproducible position and orientation for source measurement within the chamber. A calibration factor can then be derived using the calibrated source for the re-entrant chamber for a particular source and geometry. If unsealed ambient air chambers are used then temperature and pressure corrections will be required.

A long lived calibrated source (for example a caesium-137 needle) can be used as a reference source to check whether the chamber is functioning correctly. This can be compared to a short lived source (e.g. Iridium-192 or iodine-125) to establish a baseline comparison of the relative sensitivity of the system to the two sources. This baseline measurement, with the long lived source, can then be corrected for decay and be used for subsequent calibrations of the short lived source. This assumes that the energy response of the chamber does not change with time, so it is important to repeat short lived calibrations annually using new calibrated sources.

QA checks on the well-type ionization chamber should include a constancy check with a long lived reference source before each use, which checks the stability of response of the chamber with time and should be within a tolerance of ±5%[1]. IPEM Report 81[1] recommends that the energy response of the chamber be checked monthly to be within a tolerance of ±3% and calibrations should be checked annually with a tolerance of ±2%. TG56 [8] recommends additional annual checks on linearity with respect to source strength and checks on precision.

APPENDIX B

HDR source calibration

There are three methods that can be used for calibrating an HDR source:

- 'In-air' measurements with a calibrated ion chamber

- Re-entrant ionization chamber

- Solid phantom measurements with a calibrated ion chamber

The current method recommended in the UK by the joint BIR/IPSM Working Party[2] is to calibrate the HDR iridium-192 sources in air with a Farmer (0.6 cc) ionization chamber at distance of 100 mm from the source. A definitive calibration of the source should be made as part of the commissioning of new equipment and following routine source replacements. The endpoint of the calibration is to determine the reference air kerma rate of the source within an acceptable tolerance (±5%). BIR/IPSM also recommend that regular confirmatory measurements of source strength be made monthly or after any maintenance to the unit.

The traceability route for these measurements is currently via the external beam secondary standard. A thimble chamber and dosemeter are calibrated in terms of air kerma for the direct measurement of the air kerma rate (AKR). Currently national standards laboratories do not offer calibration factors for the complex spectrum of iridium-192, although in the UK, the NPL are developing this service. A calibration factor for iridium-192 has to be interpolated from calibrations traceable to the standards laboratory as close to the mean energy of iridium-192 (370 keV) as possible. Interpolation between the 250–280 kV (without a build-up cap) secondary standard air kerma factor and Co-60 factor (with a build-up cap) is probably the most practical, ensuring that the build-up cap is accounted for to provide a sound basis for the interpolation. Alternatively the BIR/IPSM[2] recommend that, until the primary calibrations are available, the calibration factor for the highest available kilovoltage quality should be used, which is heavily filtered 280 kV X-ray quality (for use with measurements made without a build-up cap). This subject has been discussed by Goetsch[19] and is well summarized in the helpful IAEA TECDOC-1274[3].

The procedure for measuring AKR is relatively straightforward and in practice a rigid device is required to hold the ionization chamber at a fixed distance from the radioactive source. The jig should provide a rigid controlled geometry, designed to support an ion chamber centrally between two plastic catheters and thus minimize dose errors due to positional uncertainties (see Fig. 12.5). The jig should be constructed to minimize scatter contribution from the jig and be positioned in the centre of the room 1.5 m away from the walls and 1 m above the floor. In practice, a small correction factor will be required to correct for scatter from the room and the jig itself. Methods for calculating this correction factor have been described by Ezzel[20] and Goetsch[19].

AKR measurements can be made in air at distances between 100 mm to 500 mm and are then converted to the reference air kerma rate at 1m. The larger the source to chamber distance, the less significant are the chamber dimensions, but the ion current may approach the limit of the measuring system. The BIR/IPSM[2] recommends 100 mm as the distance between the chamber and the source. The fluence gradient across the chamber is required to be corrected for a 0.6 cc Farmer chamber at 100 mm; correction factors are recommended in IAEA TECDOC-1274[3], and Kondo and Randolph[21]. At shorter distances, positioning is more critical due to the high dose gradients around the source, and fractions of a millimetre can give large variations in AKR. Also a larger correction factor is required for the finite size of the ionization chamber.

Goetsch *et al.*[19] demonstrated that for in-air measurements at short distances to the source, it is necessary that high energy electron contamination emitted from the source capsule is removed. The NE 2MV build-up cap maybe

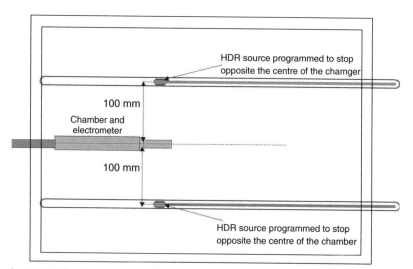

Fig. 12.5 Calibration set-up for air kerma rate measurements for a HDR Ir-192 source.

used for this purpose and provide additional build-up to provide charge particle equilibrium. The calibration factor recommended is for the ion chamber without a build-up cap, therefore a correction factor will be required for the presence of the build-up cap. BIR/IPSM[2] recommend a value of 1.017 for the 2 MV build-up cap to correct for attenuation of the chamber wall and the build-up cap.

The calibration procedure recommended by the BIR/IPSM[2] involves two parts, with each part being performed by a different physicist with a different calibrated dosemeter/chamber combination. The dosemeter/chamber combination must have been calibrated for air kerma against an NPL Secondary Standard. The Farmer 0.6 cc chamber (preferably type 2571), is fitted with an NE build up cap to act as an electron filter and is placed in the calibration room for at least 20 minutes to equilibrate with the ambient conditions. It is essential prior to any measurements that the HDR unit is functioning correctly and the temporal accuracy of the unit is within the specified tolerance. The chamber is positioned centrally between two source catheters, see Fig. 12.5, and preparatory measurements are made to ensure that the source is driven to a position opposite the chamber's effective point of measurement.

Part A: The definitive calibration

The remit of the first physicist is to measure the air kerma rate at 100 mm by exposing the source either side of the chamber. This measurement is then converted to the reference air kerma rate at 1 m by the application of the inverse square law and compared to the calibration certificate. Ionization readings are typically acquired for 300 seconds for a 100 mm source to chamber distance, ensuring that the transit dose is not included in the measurement. Consecutive readings should be within 1% of each other and the mean readings from each catheter position should be within 2% of each other. The reference air kerma rate is calculated as below and should be within ± 5% of the calibration certificate.

$$ RAKR = R \times F_c \times F_{ic} \times F_{t,p} \times F_s \times F_e \times F_g \times F_{isl} \times \frac{3600}{t} \times F_{air} $$

RAKR is measured in Gy/hr at 1 m

R   is the mean reading at 100 mm

F_c is the air kerma calibration factor for the secondary standard

F_{ic} is the intercomparison ratio between the field instrument and the secondary standard

$F_{t,p}$ is the temperature and pressure correction factor

F_s is the scatter correction factor for the room and the jig

F_e is the attenuation correction factor for the electron filter

F_g is the dose gradient factor for the 0.6 cc ion chamber at 100 mm

F_{isl} is inverse square scaling from 100 mm to 1m

t   is the time in seconds for each reading

F_{air} is the correction for air attenuation and scattering (normally assumed to be unity)

Corrections for source transit time to and from the safe have been taken as unity in the equation above, as the reading is started after the source has reached the measurement dwell position.

Part B: Confirmatory measurement

A confirmatory measurement is made by a second physicist, who will reassemble the measurement set-up, and calculate the time to deliver a specified air kerma at specified distance based on the RAKR measured in Part A. The measured value should be within ±3% of the expected value and once verified the source can be used clinically. If any of the tolerances have been exceeded then these must be investigated and reconciled before the machine can become clinical.

Alternative methods of measurement

Specially designed re-entrant ionization chambers for high strength sources are a convenient method of measuring the source strength of iridium-192 HDR sources. Jones[22], Goetsch[19] and IAEA TECDOC-1274[3] all discuss the use of re-entrant ionization chambers as part of a quality assurance programme. The re-entrant ionization chamber should be re-calibrated annually in a manner that is traceable back to a standards laboratory. For accurate definitive calibrations, the chamber should be located away from objects that might cause scatter and a perspex insert should be used to ensure positional reproducibility.

Thimble ionization chamber measurements made in solid or water phantoms are made at smaller distances between the source and chamber (between 50 to 100 mm) and therefore are more susceptible to positional uncertainties unless accurate machining is used. Ezzell[20] describes the measurement method, the main correction factors required are the dose gradient correction factor for the finite size of the chamber and a correction for the phantom not having full scatter conditions. BIR/IPSM[2] recommends that the procedure for calculating dose rate at a point in water should be to first determine the air kerma rate in water at that point and then to multiply by the ratio of the mass energy absorption coefficient for water and the mass energy transfer coefficient for air, which is taken as 1.11 for iridium-192.

APPENDIX C

HDR commissioning tests

The design of the treatment room for HDR brachytherapy including additional equipment required for the room is discussed in Chapter 3 and IPSM Report 75[23]. HDR treatment room requires walls of thickness 40–60 cm of concrete depending upon the size of the room and the energy and source strength. The room will have some form of maze type entrance and will have a protective lead door interlocked to the unit. Prior risk assessments are required before work involving ionizing radiation, which is covered in Chapter 3. A comprehensive radiation survey of the area and adjacent areas must be made prior to using the facility and should be reviewed preferably annually by appropriate monitoring or sooner if there is a change in the facilities or procedures. The survey should also include monitoring when the source is in its safe position to ensure the shielding of the source complies with regulatory requirements. It is also useful to have knowledge of the radiation distribution within the treatment room so emergency procedures can be optimized to minimize dose to staff. Commissioning tests on the equipment are performed by the purchaser's representative to ensure the equipment is ready for clinical use and to establish QA baselines. Many of the tests described previously covering source calibration, machine function tests, temporal and positional precision and applicator quality assurance will form the basis of the commissioning/annual quality assurance checks. Additional commissioning tests on the unit are indicated in Table 12.3, which will generally be performed in conjunction with commissioning checks on the treatment planning system described in Section 12.5.

Table 12.3 Additional commissioning and annual quality assurance checks

Test	Method
Radiation survey	• Perform facility survey and review, if workload or structure is revised.
Dose delivery accuracy	• Traceability to national standards laboratory
	• Verify reference air kerma rate of source and at subsequent source changes

Table 12.3 (continued) Additional commissioning and annual quality assurance checks

Test	Method
Machine function tests	• See Section 12.3.3
	• Verify all fail safe mechanisms, interlocks, monitoring, and emergency features
	• Ensure all the required equipment is available for clinical use and the emergency procedure
Positional accuracy	• Verify accuracy of all localization equipment including position of dummy markers
	• Verify source position
	• Verify transfer tube and applicator length for all scenarios
Temporal accuracy	• Absolute timer accuracy and timer linearity
	• Measure transit dose of the source
Overall system test	• Verify all steps of the brachytherapy procedure including implant reconstruction, treatment planning, data transfer, and delivery
Documentation	• Review all clinical procedures, emergency procedure and quality assurance programme for the brachytherapy facility
Training	• Ensure all personnel are trained appropriately and the emergency procedure is full understood

References

1. IPEM Report 81, Physics Aspects of Quality Control in Radiotherapy, IPEM 1999.

2. Report of a Joint Working Party of BIR and IPSM, Recommendations for Brachytherapy Dosimetry, BIR 1992.

3. IAEA (ed. H Tolli), IAEA TECDOC-1274, Calibration of photon and beta ray sources in brachytherapy, IAEA, 2002.

4. Jones CH (2001). Calibration of sources. In: Joslin CAF, Flynn A, Hall EJ (ed.). *Principles and practice of brachytherapy.* London: Arnold.

5. Jones CH (2001). Quality assurance in high dose rate afterloading. In: Joslin CAF, Flynn A, Hall EJ (ed.). *Principles and practice of brachytherapy.* London: Arnold.

6. Slessinger ED. Quality assurance in low dose rate afterloading. In: Joslin CAF, Flynn A, Hall EJ (ed.). *Principles and practice of brachytherapy.* London: Arnold.

7. Glasgow GP Bourland JD, Grogsby PW, Meli JA, Weaver KA (1993). Remote afterloading technology, AAPM 41, report of the AAPM Radiation Therapy Committee Task Group No. 41.

8. Nath R, Anderson LL, Meli JA, *et al.* (1997). Code of practice for brachytherapy physics. Report of the AAPM Radiation Therapy Committee Task Group No. 56. *Medical Physics* **24**(10):1557–98.

9. Fraass B, Doppke K, Hunt M, *et al.* (1998). Quality Assurance for Clinical Radiotherapy Treatment Planning: Report of the AAPM Radiation Therapy Committee Task Group No. 53. *Medical Physics* **25**(10):1773-1829.

10. ICRU 38, Dose and Volume Specification for Reporting Intracavitary Therapy in Gynaecology, International Commission on Radiation Units and Measurements, 1985.

11. Wong TPY, Fernando W, Johnston PN, Bubb IF (2001). Transit dose of an Ir-192 high dose rate brachytherapy stepping source, *Phys. Med. Biol,* **46**(12):323–331.

12. Sahoo N (2001). Measurement of transit time of a remote afterloading high dose rate brachytherapy source. *Medical Physics* **28**:1786–1790.

13. Aukett RJ (1991). A technique for the local measurement of air kerma rate from small caesium-137 sources, *British Journal of Radiology* **64**:918–922.

14. Williamson JF. Practical quality assurance in low dose rate brachytherapy. In: Strakschall, Horton J (ed.). *Quality Assurance in Radiotherapy Physics: Proceedings for an American College of Medical Physics Symposium,* pp. 139-82 Medical Physics Publishing Company.

15. Van Dyk J, Barnett J, Cygler J, Shragge P (1993). Commissioning and quality assurance of treatment planning computers, *International Journal of Radiation Oncology, Biology and Physics* **26**:261–73.

16. Nath R, Anderson LL, Luxton G, *et al.* (1995). Dosimetry of interstitial brachytherapy sources: recommendations of the AAPM Radiation Therapy Committee Task Group No 43, *Medical Physics* **22**:209–34.

17. http://www.estroweb.org/ESTRO/upload/EQUALdocs/GeomRecApplicationForm.pdf

18. Williamson J F, Morin RL, Khan FMl (1983). Dose calibrator response to brachytherapy sources: as Monte Carlo and analytic evaluation. *Medical Physics* **10**(20):135–40.

19. Goetsch SJ, Attix FH, Pearson DW, Thomadsen BR (1991). Calibration of Ir-192 High Dose Rate Afterloading Systems. *Medical Physics* **18**:462–7.

20. Ezzel G (1989). Evaluation of calibration techniques for the microselectron HDR. In: *Brachytherapy,* 2nd edn. pp 61–69. Mould RF Leersum, The Netherlands, Nucletron.

21. Kondo S, Randolph ML (1960). Effect of finite size of ionisation chambers on measurement of small photon sources. *Radiation Research* **13**:37–60.

22. Jones CH (1995). HDR microselectron quality assurance studies using a well-type ion chamber. *Phys Med Biol* **40**:95–101.

23. IPSM Report 75, The Design of Radiotherapy Treatment Rooms, IPSM 1997 (reprinted IPEM 2002).

Index